Innovations of Green Process Engineering for Sustainable Energy and Environment 2016

Topical Conference at the 2016 AIChE Annual Meeting

San Francisco, California, USA
13 - 18 November 2016

ISBN: 978-1-5108-3428-6

Printed from e-media with permission by:

Curran Associates, Inc.
57 Morehouse Lane
Red Hook, NY 12571

Some format issues inherent in the e-media version may also appear in this print version.

Printed by Curran Associates, Inc. (2017)

AIChE
120 Wall Street, FL 23
New York, NY 10005-4020

Phone: (800) 242-4363
Fax: (203) 775-5177

www.aiche.org

Additional copies of this publication are available from:

Curran Associates, Inc.
57 Morehouse Lane
Red Hook, NY 12571 USA
Phone: 845-758-0400
Fax: 845-758-2633
Email: curran@proceedings.com
Web: www.proceedings.com

TABLE OF CONTENTS

460933 Reutilization of Coal Fly Ash for the Production of Highly Beneficial Products

Abstract:

Coal fly ash (CFA), generated during the combustion of coal in power generation plant, is an industrial waste that continues to increase as a result of the growth in power demand. This large amount of industrial waste can cause environmental and human health problems if it is not treated properly. Research on the potential reuse of CFA has therefore been undertaken intensively worldwide. From those perspectives, this work aims to reduce the problems related to the disposal of CFA by converting them into beneficial products, including zeolite, geopolymer and fertilizer.

Zeolites: Hydrothermal alkaline reaction was employed for converting CFA to zeolites. Two different types of alkali (i.e. NaOH and KOH) were used as activating agents. Different Si/Al ratio, time and temperature used for the hydrothermal reaction were investigated for their effects on the type and amount of zeolites synthesised. It was found that, by using NaOH as activating agent, a mixture of Na-A and hydroxy-sodalite was mostly formed. On the other hand, a different mixture of merlinoite and perlialite was formed when KOH was used as the activating agent. The obtained zeolites were then employed for dye removal from wastewater and it was found that zeolite obtained from NaOH has the highest adsorption capability of 16.26 mg/g.

Geopolymer: In order to produce CFA-based geopolymer, CFA was firstly treated with the aqueous solution of Na_2SiO_3 and NaOH. Concentration of Na_2SiO_3, SiO_2/Na_2O mole ratio, ash/alkaline solution mass ratio, temperature and curing time were varied to study their effects on heat resistance and compressive strength of geopolymers. It was found that the compressive strength of geopolymer increases with Na_2SiO_3 concentration, temperature and curing time, whereas heat resistance of geopolymer is insignificantly affected by varying those parameters (dry weight loss between 2.5 and 4.9%). Moreover, it was found that geopolymers produced have higher compressive strength (1115 MPa) and better heat resistance as compared to the hardened Ordinary Portland Cement (OPC) (400 MPa).

Fertilizer: CFA consists of all the elements present in soil except organic carbon and nitrogen, making it becomes a valuable raw material for producing fertilizer. In order to produce fertilizer, CFA was mixed with another two components: (i) wood compost, which contains significant amounts of basic nutrients for the plants i.e. carbon, hydrogen oxygen nitrogen and sulfur; and (ii) biochar, which can provide a physical domain for bacteria and fungi, including beneficial mycorrhizal species and also enhance the water-holding capacity of the soil. From the ICP analysis, the fertilizer produced contains significant amounts of useful nutrients for plants, including phosphorus and potassium as primary macronutrients; calcium and magnesium as secondary macronutrients; cobalt, copper, iron, manganese, sodium, and zinc as micronutrients. Moreover, the final product also contains a significant amount of silica (~18.9 wt. %). Last but not least, it was found that the fertilizer obtained can significantly enhance the quality of soil.

468869 Characterizing Catalytic Waste Gasification - Effects of Substrate and Catalyst

Abstract:

This research is primarily aimed to extend proof-of-concept gasification experiments as a suitable alternative for waste management and sustainable technology. While initially intended to make more efficient use of resources in space exploration missions by producing high value reaction products from normally unused or wasted materials, the same principles can be applied to municipal waste as well.

The experiments performed examine gasification dynamics for different substrates promoted by different catalysts. It utilizes a high pressure batch scale reactor which allows examining controlled multi-phase catalytic isothermal oxidation of solid and melting substrates in the presence high purity air. The reaction time was varied between 2 and 24 hours and a preliminary overall kinetic model was formulated based on conversion, selectivity, and yield determinations. The studied substrates, Polyethylene and Cellulose, are both long chain organic polymers, and make up a substantial portion of both space and municipal waste composition. Although similar in nature, these substrates exhibit marked differences as it pertains to gasification, and were therefore selected as model substrates. Namely, cellulose contains oxygen as part of its molecular structure, while polyethylene does not. Since the technology investigated is aimed at working to relatively low temperatures (below 350 ^{o}C), cellulose remains as a solid particulate at the investigated reaction temperature while polyethylene melts.

Following the reaction phase, both the liquid and solid reaction residues, along with the gaseous products, are collected for characterization and analysis. Gas quantification is performed using a Gas Chromatograph. While thermal characterization (DSC) and Scanning Electron Microscopy (SEM) are used to characterize solid and liquid residuals.

Preliminary results suggest that catalytic gasification provides potential control for the liquid-phase oxidations. Under appropriate conditions, oxidation was found to follow the incomplete oxidation pathway which resulted in significant yields of hydrogen and methane, which are potentially high value products as they can be used for energy recovery.

Comparisons between the gasification rates for the two substrates are drawn. Similarly, comparative studies of different catalysts are also presented. The results are used in conjunction with a kinetic model for the gas-phase to formulate a methodological model for the gasification of solid waste. The results will be presented along with a preliminary extension for a continuous gasification process.

459628 Experimental and Numerical Study on Combining Anaerobic Digestion and Gasification for Potential Energy Recovery from Biomass and Sewage Sludge

Abstract:

The purpose of this study is to improve energy recovery and realize full utilization of municipal solid wastes (MSWs). This paper proposes a novel concept of a two-stage MSWs disposal process that combines the anaerobic digestion and biomass gasification. Both experimental and numerical studies have been conducted to investigate the performance of an integrated system for the energy recovery from MSWs based on anaerobic digestion and gasification. The experiments are performed at two stages. At the first stage, anaerobic digestion of MSWs, such as grass, leaves and animal manure, is conducted by mixing with anaerobic sludge. At the second stage, the digested solid residue is co-gasified with woody biomass for syngas production. The effect of inoculum to substrate ratio on anaerobic digestion and wood chips to digested solid residue ratio on gasification performance are investigated respectively. A combined kinetic model is then built for prediction and optimization of biogas and syngas yields in this two-stage process. The deviation between the numerical and experimental results obtained was lower than 10%. Both experimental and numerical results show that full utilization of MSWs can be realized to produce energy through anaerobic digestion followed by gasification process.

461746 Hydrothermal Carbonization of Organic Fraction of Municipal Solid Waste and Corresponding Digestate Via Anaerobically Digestion

Abstract:

Steam autoclaving is a technology that has been around since the 80â€™s and can efficiently divide the biogenic and non-biogenic fractions of Municipal Solid Waste (MSW) for best re-use and upgrading. Autoclaving eliminates hazardous hand sorting of raw garbage and in the process of separating the pulp fibers, sterilizes the output and allows for sequestration of pollutants that could otherwise leach into the groundwater or the atmosphere over time. The organic fraction is recovered by trommel screen (accepts <3/8â€) and is comprised of a moist pulp matrix embedded with ash and debris. The food waste from MSW is solubilized and absorbed onto the pulp matrix, creating a unique feedstock for further thermochemical processing (e.g., hydrothermal carbonization, HTC) biological processing (e.g., anaerobic digestion, AD), or a combination of both.

Hydrothermal carbonization (also called thermal hydrolysis, or wet torrefaction) is a treatment process which converts moist feedstocks into homogenized, carbon rich, and energy dense solid fuel, called hydrochar. One of the main advantages of HTC compared to other thermochemical treatment processes is the use of moisture as reaction medium and reactant, thereby precluding the need for drying prior to treatment. Thermodynamic properties of water change greatly in the subcritical region from 180-280 °C, and as a result, subcritical water behaves as a non-polar solvent and mild acid and base catalyst simultaneously. When subjected to HTC, biomass undergoes rapid hydrolysis and other reactions, releasing oxygen-containing volatiles and producing a highly hydrophobic hydrochar. Aqueous products typically include organic acids, sugars, and furans.

In this study, the autoclaved organic fraction of municipal solid waste pulp (OFMSW) and the digestate from OFMSW pulp after AD were processed by HTC at 200, 250, and 300 °C for 30 min and 2 h. The focus of this work was to evaluate the potential for producing an energy-dense, stable, solid hydrochar by HTC from either OFMSW or its digestate. The fate of inorganics, especially the heavy metals in hydrochar is discussed. In addition, the aqueous products were analyzed to identify the chemicals produced and the chemical changes during HTC. A carbon balance was performed from the solid and liquid phase analyses. Moreover, two conceptual MSW treatments consisting of autoclaving with or without AD, but including HTC were proposed and mass and energy balances were performed. The proximate, ultimate, FTIR, and fiber analyses show that the fuel value of the hydrochar increases with increasing HTC temperature. Inorganics have a strong propensity for the hydrochar matrix, and are found only in small amounts in the aqueous phase. The fuel content is as high as 44.7 MJ kg^{-1} (dry ash-free basis) when HTC digestate was treated at 300 °C for 2h. A cascaded design including AD and HTC yields a higher net energy production than HTC or AD alone.

461951 Minimizing Municipal Solid Waste Accumulation through the Application of Appropriate Technology

Abstract:

Along with population and urbanization in developing countries, the amount of municipal solid waste generated is also increasing. Although programs and initiatives to recycle and manage waste can often be found in the major population centers, these technologies are slow to spread to or are not yet present in the rural areas. Heavily populated urban slums are also lacking in the infrastructure needed to collect and manage trash, particularly plastic packaging. Solving the problems caused by this accumulation of municipal solid waste (MSW) derived plastic is a global challenge. Plastic waste accumulation in landfills and oceans is detrimental to ecosystems as well as public health. In developing regions particularly, where MSW handling infrastructure is limited or unavailable, plastic is often simply left on the ground where it eventually makes it way to the ocean. Because these sources are widespread, the solutions to dealing with this plastic must also be widely dispersed.

Fortunately, many plastics can easily be converted into a fuel oil that can be directly used as a substitute for petroleum derived kerosene or diesel. Plastics commonly found in MSW such as High Density Polyethylene (HDPE), Low Density Polyethylene (LDPE), Polypropylene (PP), and Polystyrene (PS) are all viable raw materials for this process. Plastics such as polyethylene terephthalate (PET) and poly vinyl chloride (PVC) are unfortunately not suitable for this process, due to the presence of oxygen and chlorine respectively in the polymers. The chemistry of converting plastic waste to a hydrocarbon fuel oil is simple and well established. Since, the molecules of these plastics are only made up of carbon-hydrogen chains, when thermally heated to temperatures of approximately 430°C (Wong, et al., 2015), the hydrocarbon chains break, decomposing the polymer, and yielding a hydrocarbon gas, which is then condensed to obtain the fuel oil product. However, industrial processes for carrying out this chemistry are typically not available in the developing regions that are the source of much of the plastic that ends up in the ecosystem. To address this problem, solutions based on combining the principles of appropriate technology with green process engineering are needed.

Appropriate technology (AT) is a design philosophy first described in 1973 by E.F. Schumacher in his book *Small is Beautiful*. According to Hazeltine and Bull in their 1999 book "*Appropriate Technology: Tools, Choices, and Implications*", it refers to "technological choices and applications that are small scale, decentralized, labor-intensive, energy efficient, environmentally sound, and locally controlled" (Hazeltine, 1999). AT is simply technology suitable for a specific region, designed to meet specific needs of certain individuals or communities. It differs from industrial scale processes in the sense that AT is specifically intended to be used in rural or underdeveloped regions. AT is low-cost, small-scale, easy to construct, and can be operated by individuals with limited formal technical education. The purpose of AT is not to reproduce industrial technology on a small scale; instead, it is to design specific solutions based upon the resources available in a given region or for a given community.

It is our assertion that appropriate technology is the most viable option for solving the global problem of waste plastic accumulation in developing regions. Therefore, we will make the case that through the utilization of a low-cost easy to use processor for converting waste plastic to fuel oil, a sustainable solution for reducing the volume of waste plastic entering the ecosystem can be achieved. We will address the challenges for implementing such a strategy and discuss how these challenges can be met.

460695 Struvite Precipitation for Nitrogen Removal in Horse Manure for Gasification: Experimental Study and Economic Analysis

Abstract:

Animal manure is one of the most common kinds of waste in farming areas. Instead of direct disposing or burning, gasification is a more promising and cleaner way to harvest the energy from waste and reduce pollution at the same time. From previous study, addition of horse manure and chicken manure to woodchips could improve syngas production up to 40%. However, the nitrogen content inside the manure, taking up around 2-3 wt% in the sample, may course environmental or explosive concern for gasification. On the other hand, retrieving nitrogen from animal manure can be a significant source as fertilizers. From FT-IR tests, nitrogen was found to be in the form of ammonia mainly in the manure samples. Thus, struvite precipitation is considered as an excellent approach to remove ammonia and produce struvite at the same time. In this work, horse manure from Singapore Turf Club was studied. Elemental analysis, FT-IR and TGA tests for the manure samples were conducted as feedstock characterization. Firstly, ammonia was leached out from manure samples using water at different ratios. After leaching, ammonia concentration could reach around 100 mg/L. For precipitation process, optimal experimental conditions, such as molar ratios of Mg^{2+} and PO_4^{3-} sources, mixing time and pH, were studied. Performance of different kinds of Mg^{2+} and PO_4^{3-} sources were compared. In the end, economic analysis of struvite precipitation process was also conducted to find the optimal solution for the whole process. The objective of this work is to find out the optimal experimental conditions for fully utilization of manure waste in a most cost-effective and high-yield way.

454626 Highly Selective Conversion of Biogenic Waste Materials to "Biogenic Formic Acid" As a Green Platform Chemical

Abstract:

The selective oxidation of complex, water-insoluble and wet waste biomasses from second and third generation to formic acid including effective catalyst recycling is reported. Additionally, the relevance and limits of potential contaminants are illustrated by different experimental approaches. By using a very robust homogeneous polyoxometalate catalyst in aqueous solution, molecular oxygen as oxidant and an acid as solubilizer, it is possible to convert different lignocellulosic and algae feedstock into formic acid and pure carbon dioxide. The applied green oxidation system benefits from its low reaction temperature (below 100 °C) and its very selective nature. The upcoming requirements for sustainable production of chemicals will lead to an increasing substitution of fossil fuels by biorenewable feedstock. This is mainly due to the fact that almost all use of hydrocarbon fossil fuels in production processes is directly or indirectly linked to carbon dioxide (CO_2) emissions that contribute to global warming. However, biomass conversion is usually characterized by complex feedstock compositions leading to sophisticated processes with many parallel and consecutive reactions involved. This often results in low yields of the designated products. Consequently, only few processes for directly converting biomass into value-added chemicals have been established in academic research. The high oxygen content of biomass requires typically a number of reductive conversion steps in biomass valorization sequences leading almost unavoidably to drawbacks like formation of tar or gluey byproducts. Moreover, the ultimate goal of producing valuable chemicals in high yields by direct conversion of complex biomasses, such as e.g. lignocellulosic substrates or algae is still far ahead. Major difficulties arise from the fact that such substrates are not well soluble in conventional solvents and are very resistant to chemical and biological transformations.

Other important considerations for the chemical use of biogenic feedstocks arise from conflicts with the use of biomass as food and animal nutrition. To characterize the potential for such conflicts, the broad range of biogenic feedstock conversion processes has been classified into three generations: The first generation of biomass conversion processes deals with feedstock that is in direct conflict with the supply of food. Examples are sugars, starch and vegetable oils. Conversion processes that utilize lignocellulosic biomass like wood or herbage do not use feedstock that is in direct conflict with food supply. However, these biomasses grow on farmland and are therefore in indirect conflict with food supply. The use of microorganisms, such as algae, to produce fuels or chemicals is denoted as third generation biomass conversion processes. These processes are preferred as future biomass conversion processes as no direct or indirect conflict with food and nutrition applications arises from a large scale, industrial realization.

Formic acid (FA) is a basic chemical that is widely used in chemical, textile, leather, pharmaceutical, rubber and other industries. Recently, FA is discussed in addition as fuel for direct FA fuel cells, and as hydrogen storage material. FA can be easily and selectively decomposed to hydrogen and CO_2through metal catalysed processes under very mild conditions. In addition, thermal decomposition of FA (above 373 K) forms CO and water. Thus, FA can be regarded as a liquid syngas equivalent and conversion of biomass to FA can be regarded as up-grading an abundant natural resources into a green energy and synthesis equivalent.

Oxidative conversion of wet biomass to formic acid using homogeneous POM catalysts has been developed by our group in the recent five years (OxFA-process). The concept essentially overcomes all major problems of classical biomass gasification or reforming processes. The interesting difference of the OxFA route compared to reductive biomass valorization approaches is that all thermally induced formation of solid or gluey by-products is completely avoided. Moreover, all heteroatoms in the biomass are converted into their highest oxidized, water-soluble forms (e.g. sulfates, nitrates) in contrast to the formation of H_2S and ammonia encountered under reductive conditions.

The OxFA-process is mildly exothermic and operates under mild temperature conditions of typically below 373 K. As water is applied as solvent, biomass of different origin, composition and humidity can be applied without drying. Formic acid is produced as the only product in the liquid phase avoiding tedious liquid phase separation processes. The only side-product is pure carbon dioxide that forms in the gas phase and can be re-used for the production of new biomass substrate (e.g. for the production of algae in an algae farm). As the OxFA process converts a very broad range of biogenic raw materials into only two products that separate nicely into gas and liquid phase its simplicity and robustness are clear advantages compared to other biomass valorization technologies. The OxFA process applies homogeneous polyoxometalate (POM) complexes as redox catalysts. POMs are characterized by strong Brønsted acidity, high proton mobility, fast multi-electron transfer, high solubility in various solvents and excellent resistance against hydrolytic or oxidative degradations in solution. The named properties make this class of catalysts very attractive for the oxidative conversion of biomass in aqueous media. In particular, their tunable redox potential and acidity offers enormous potential for their future use in oxidation chemistry. To increase the efficiency of the OxFA process for the conversion of water-insoluble, complex biomass, we recently introduced the use of sulfonic acids as additives. The latter have been found to act as efficient depolymerization catalysts, solubilizers and promotors. These findings have paved the way for the oxidative processing of e.g. lignocellulosic biomass, a class of substrates that is notoriously difficult to convert in aqueous phase reforming.

Oxidation experiments for expanding the substrate scope

In this contribution, we significantly expand on the scope of biogenic substrates that can be processed with the OxFA-process. Therefore, we first studied a range of second generation feedstocks like cane trash, pomace, fruit pulp, bark from hard and softwood, leafs, straw, grass clippings and aerated material. As representatives for third generation biomasses, we tested green algae like *chlorella* and *ulva lactuca*, brown algae like *ascophyllum nodosom*, blue-green algae like *spirulina* or *cyanobacter* and *chondrus crispus*as red algae. To enlarge the list of feedstocks even to the treatment of problematic and contaminated biological waste materials, we also applied effluent sludge and deinking sludge from a paper manufacturing plant under typical OxFA process conditions. Additionally, we used railway sleepers and beech condensates as substrates for the process. The effluent sludge contained about 80 wt% organic materials with a high nitrogen content of about 10 wt%. This substrate also contained large amounts (ca. 10 wt% in total) of inorganic materials and salts like calcium oxide and different phosphates. The applied deinking sludge was of even more inhomogeneous composition. Besides 35 wt% of organic material, it contained many different heavy metal salts like copper, lead, cadmium and nickel compounds. Furthermore, this substrate contained highly volatile halogenated hydrocarbons, like chloroethane and vinyl chloride as well as polycyclic and polychloric hydrocarbons. The applied railway sleepers were characterized by a high content of aromatic compounds. The applied beech condensate was also rich in aromatic compounds.

Based on our previous findings, we applied for the screening experiments of the named new substrates molecular oxygen as the oxidant, the polyoxometalate $H_5[PV_2Mo_{10}O_{40}]$ (HPA-2) in aqueous solution as a catalyst and para-toluenesulfonic acid (TSA) as an additive and solubilizer. The conversion of all substrates was performed in a 600 mL Hastelloy autoclave with a gas entrainment impeller and additional cooling to deal with the exothermicity of the reaction. Effects of different liquid volumes in the reactor (100 mL of reaction mixture in 600 mL of reactor volume) can be excluded as the reaction takes place only in the liquid phase where the oxygen is entrained by the impeller. In particular, there is no further reaction in the gas phase. All biogenic feedstocks were processed for 24 hours at 90 °C under 30 bar oxygen pressure. The initial composition of the reaction mixture in each oxidation experiment was 100 g of water as a solvent, 1.74 g (0.9 mmol) of HPA-2 polyoxometalate as a catalyst, 1.90 g (11 mmol) of TSA as additive/solubilizer and 3.30 g of the substrate under investigation. To enable the calculation of yields based on the comparable basis of carbon atoms available in the substrate, we performed an elemental analysis of each substrate prior to our catalytic experiments (results see Table 1).

The results in Table 1 clearly show that the applied homogeneous oxidation system is able to convert all applied 2nd and 3rd generation biomasses to predominantly formic acid and carbon dioxide. Under the applied conditions no CO formation from FA decomposition and no catalytic FA decomposition to $CO_2$and hydrogen were detected (see Table 1, entry 19). The yield in FA was found to depend strongly on the structure of the substrate and the number of oxygen-functionalities in the carbon-framework. Among the tested new substrates, the best raw material for producing formic acid was found to be pomace with 55 % FA-yield (Entry 1) followed by cane trash with 49 % FA-yield (Entry 2) and fruit pulp with 45 % FA-yield (Entry 3). Note that all these latter three origin from commercial and industrial biomass conversion processes making the here presented technology an interesting add-on to widely practiced processes of the food industry.

Table 1: Oxidative conversion of complex, water-insoluble biomass of 2nd and 3rd generation in presence of the additive TSA using a HPA-2 POM-catalyst.

Entry	Substrate	Molecular composition[a]	Combined Yield[b] FA+CO_2 [%]	FA-Yield[b] [%]	FA:CO_2-selectivity[b] [%]
1	Pomace	$C_{0.24}H_{1.59}O_1$	97.3	54.6	

56:44

2

Cane trash

$C_{0.92}H_{2.04}O_1$

97.5

48.9

50:50

3

Fruit pulp

$C_{1.00}H_{1.92}O_1$

99.0

45.3

46:54

4

Spruce chips

$C_{1.30}H_{1.98}O_1$

89.5

34.6

39:61

5

Poplar splint

$C_{1.32}H_{1.98}O_1$

66.0

30.8

47:53

6

Grass clippings

$C_{1.17}H_{2.05}O_1$

74.6

29.9

40:60

7

Chondrus crispus

$C_{0.71}H_{1.30}O_1$

48.3

21.7

45:55

8

Chlorella

$C_{1.69}H_{2.92}O_1$

53.7

21.6

40:60

9

Oak bark

$C_{1.21}H_{1.87}O_1$

56.0

21.1

38:62

10

Willow bark

$C_{1.34}H_{2.15}O_1$

59.5

16.1

27:73

11

Cyanobacter

$C_{0.72}H_{1.32}O_1$

27.4

15.8

58:42

12

Spirulina

$C_{1.70}H_{2.99}O_1$

40.6

15.7

39:61

13

Birch bark

$C_{1.39}H_{2.29}O_1$

41.0

15.5

38:62

14

Ascophyllum nodosum

$C_{0.82}H_{1.42}O_1$

29.2

14.8

51:49

15

Straw

$C_{1.04}H_{1.62}O_1$

53.1

11.9

22:78

16

Nettle leafs

$C_{1.10}H_{1.98}O_1$

54.2

11.0

20:80

17

Ulva lactuca

$C_{0.75}H_{1.53}O_1$

16.4

7.6

46:54

18

TSA

$C_7H_8O_3S$

-

-

-

19

FA

CH_2O_2

-

-

-

20

Neat pomace[c]

$C_{0.24}H_{1.59}O_1$

52.0

12.5

24:76

21

Neat cyanobacter[c]

$C_{0.72}H_{1.32}O_1$

14.8

4.3

29:71

Reaction conditions: 3.3 g substrate, 1.9 g (11 mmol) additive and 1.74 g (0.9 mmol) catalyst dissolved in 100.0 mL H_2O, 30 bar O_2, 90 °C, 24 h, 1000 rpm; [a] Determined via C, H, N, S elemental analysis; [b] Yield and selectivity determined by means of ^{1}H-NMR using benzene as an external standard according to n (FA)/n(C-atoms feedstock); [c] Reference experiments without TSA additive.

Interestingly, also substrates from the third generation like red algae (Chondrus crispus) or green algae (Chlorella) gave surprisingly high FA-yields. Note, that for these substrates the selected reaction time of 24 hours was definitely too short, so not all the substrate was converted to FA and CO_2. From our experience, it can be expected that prolonged reaction times would lead to a final FA yield for these substances close to the FA selectivity obtained after 24 h reaction time. These experiments could also confirm the very positive effect of the toluenesulfonic acid (TSA) additive and solubilizer on the POM-catalyzed biomass oxidation. Experimental evidence is provided by the reactions with and without the TSA additive (compare Table 1, entries 1 and 11 against entries 20 and 21). For pomace, the experiment without TSA (Entry 20) showed only very little FA formation (Y (FA) =12.5 %) compared to a FA-yield of 54.6 % (entry 1) with TSA for otherwise identical conditions. For cyanobacter, the experiment without TSA (entry 21) yielded only 4.3 % FA after 24 h reaction time, in contrast to the experiment with the TSA additive in entry 11 (15.8 % FA-yield). Based on these results, it is obvious that the TSA additive does not only act as a solubilizer but also as reaction accelerator and selectivity enhancer. The latter can be explained by the solubilizing effect of the TSA additive giving the homogeneous POM catalyst more access to the functional groups within the carbon-framework. As these groups are normally carbonyl- or hydroxy-groups this leads to an increase of the FA-yield based on our previous findings with different funtionalized C_2-model compounds. These results are in good agreement with previous findings concerning the suitability of the TSA additive for the transformation of weekly water soluble biomass substrates using homogeneous POM catalysts in water. A baseline experiment using pure TSA as the substrate showed again no oxidation neither to FA nor to CO_2(see Table 1, entry 18).

In an additional set of experiments, we tested conversion of chemically contaminated substrates in the POM catalyzed oxidation. For this purpose, we processed effluent sludge, railway sleeper, beech condensate and deinking sludge under typical OxFA conditions. In these experiments, the initial composition of the reaction mixtures contained 100 g of water as a solvent, 1.74 g (0.9 mmol) of HPA-2 as a catalyst and between 1.85 g and 10 g (100 mmol) of the substrate dependent on its water content. The additive TSA was not used in these experiments as the unknown composition of the applied â€œreal lifeâ€ substrates was expected to open additional but unknown reaction pathways for TSA (e.g. salt formation with metal ions). This would lead to consecutive reactions and uncontrollable loss of the additive which would have made interpretation of the obtained results rather difficult. For all experiments of this series the reaction temperature was kept at 90 °C in order to prevent thermal FA decomposition. The oxidation reactions were carried out at 30 bar oxygen atmosphere with stirring of 1000 rpm for 24 h reaction time. The obtained results are summarized in Table 2.

Table 2: Oxidative conversion of chemically contaminated biomass substrates with the HPA-2 catalyst.

Entry	Substrate	Molecular composition[a]	Combined Yield[b] FA+CO_2 [%]	FA-Yield[b] [%]	FA:CO_2-Selectivity[b] [%]
1	Effluent sludge	$C_{6.31}H_{11.35}O_{6.63}N_1$	46.1	15.7	34:66
2	Railway sleeper	$C_{1.25}H_{1.90}O_1$	39.5	11.2	29:71
3	Beech condensate	$C_{1.21}H_{1.86}O_1$	2.5	1.0	41:59
4	Deinking sludge				

$C_{0.38}H_{0.47}O_1$

-

-

-

Reaction conditions: 100 mmol substrate (1.85 g for entry 1; 5 g for entry 2; 10 g for entry 3 and 5.35 g for entry 4) depending on molecular weight; 1.74 g (0.9 mmol) catalyst dissolved in 100.0 mL H_2O; 30 bar O_2, 90 °C, 24 h, 1000 rpm. [a] Determined via C, H, N, S elemental analysis; [b] Yield and selectivity determined by means of ^1H-NMR using benzene as an external standard according to n(FA)/n(C-atoms feedstock).

It is obvious that the HPA-2 catalyst also works with contaminated feedstock and is highly robust against many impurities. Thus effluent sludge with the highest amount of organic and oxidisable matter yields nearly 16 % FA at a combined FA + CO_2 yield of 46 % after 24 h reaction time. The high nitrogen content within the sludge from a wastewater treatment plant has a negative effect on the conversion rate but does not prevent the oxidation reaction. Also for railway sleepers containing high amounts of aromatic compounds, the oxidation to FA and CO_2 is observed. A FA + CO_2combined yield of 40 % is observed after 24 h resulting with FA selectivity of 29 %. In contrast to the latter two substrates, highly diluted beech condensate and deinking sludge show almost no oxidative conversion after 24 h reaction time.

Defined poisoning experiments for the POM catalyst

In order to elucidate the complete lack of reactivity with the deinking sludge substrate we performed a couple of poisoning experiments to learn more about the sensitivities of the applied POM oxidation system. Our first hypothesis was that the heavy metal content in the applied deinking sludge (Palm GmbH & Co. KG paper manufacturing company) could be responsible for the observed catalyst deactivation. Such heavy-metal cations are expected to form water-insoluble and inactive complexes with the anion of the polyoxometalate catalyst. To confirm this hypothesis, we added on-purpose water-soluble salts of higher transition metals like copper, lead, cadmium and nickel as contaminants to a glucose oxidation experiment. Moreover, similar contamination experiments were performed with other typical components of deinking sludge, such as e.g. gypsum, kaoline, TiO_2 or ZrO_2. For these experiments, 5 g (25 mmol) of glucose as a water-soluble substrate and 1.74 g (0.9 mmol) of the catalyst dissolved in 100 mL water, together with 8 mmol (0.63-2.18 g) of the impurity were processed under 90 °C, 30 bar oxygen with 1000 rpm for 6.5 h in a 600 mL high-pressure autoclave. All results are displayed in Table 3; the baseline result for a non-contaminated glucose oxidation experiment is shown as entry 9.

Table 3: HPA-2 catalysed glucose oxidation - defined poisoning experiments with typical impurities from paper manufacturing to elucidate the lack of catalytic activity for deinking sludge as substrate.

No.	Impurity	Potential source of origin	Combined Yield[a] FA+CO_2 [%]	FA-Yield[a] [%]	FA:CO_2-Selectivity[a] [%]

1

$CuSO_4$

printerâ€™s ink

39.8

-

0:100

2

$CdCl_2$

printerâ€™s ink

87.6

44.4

51:49

3

$NiCl_2$

printerâ€™s ink

n.d.

n.d.

n.d.

4

$PbCl_2$

printerâ€™s ink

6.2

4.4

71:29

5

Gypsum

filler

89.9

47.5

53:47

6

Kaoline

filler

93.7

49.2

53:47

7

TiO_2

filler

72.3

44.4

61:39

8

ZrO_2

paper varnish

83.0

38.8

47:53

9

-

-

100

46.2

46:54

Reaction conditions: 5 g (25 mmol) glucose as a substrate; 1.74 g (0.9 mmol) HPA-2 as a catalyst dissolved in 100.0 mL H_2O; 8 mmol (0.63-2.18 g) of the impurity depending on molecular weight; 30 bar O_2, 90 °C, 6.5 h, 1000 rpm. [a] Yield and selectivity determined by means of ^{1}H-NMR using benzene as an external standard according to n(FA)/n(C-atoms feedstock).

The results in Table 3 clearly indicate a strong influence of Pb- and Cu-salts on the catalytic activity for glucose oxidation. Other heavy metal ions, like Cd^{2+}, appear to be of little effect. Interestingly, the presence of copper ions does completely inhibit the formation of FA from glucose oxidation, only CO_2 is detected as a product. This goes along with the observation that copper ions form water-insoluble complexes with the catalyst heteropolyanion in form of small blue particles at the bottom of the reaction vessel. Unfortunately, the oxidation result with the added $NiCl_2$ could not be quantified by the usual NMR-method due to the strong deformation of the spectra due to the paramagnetic nature of Ni (II) in the sample. Nevertheless, also for this case, the formation of FA could be qualitatively confirmed by the help of ^{13}C-NMR. In contrast to the experiments with Cu, no formation of water-insoluble crystals was observed in the case of $NiCl_2$ addition. Finally, the addition of Pb (II) species to the aqueous glucose solution resulted in an immediate color change from orange to a blue-green solution. The ^{51}V-NMR spectra of the obtained solution strongly indicate complexation of the heteropolyanion with the Pb^{2+} cations. In addition, the formation of yellow crystals at the bottom of the reaction vessel was observed thus removing the active catalyst from solution. In general, ^{51}V-NMR spectra of the reaction solutions containing late transition-metal cations show significant changes after addition of Cu-, Ni- and Pb-salts. In contrast, the ^{51}V-NMR spectra with added fillers do not show significant changes of the polyoxometalate catalyst. Addition of these water-insoluble fillers, like gypsum, kaoline, titanium dioxide and zirconia, resulted in an increase of the viscosity of the reaction mixture; however, this had very little effect on the observed glucose oxidation reaction.

Our contribution gives strong support to the scientific and economic attractiveness of homogeneously catalyzed biomass oxidation to formic acid by polyoxometalate catalysts. We could expand the substrate scope for the reaction considerably by converting a very wide range of substrates including many complex water-insoluble biogenic feeds and also contaminated materials. While waste from the food industry, effluent sludge and even railway sleeper was found to be effectively converted under typical OxFA reaction conditions (90 °C, 30 bar O_2, 24 h reaction time), a clear and so far undetected limitation has been found for metal ion containing feedstocks. Whenever contaminated biological feed with metal ions forming stable water-insoluble complexes with the catalytically active polyoxometalate ion were present in the applied feedstock, the reactivity of the oxidation system was strongly reduced or almost completely eliminated. The favorable role of the TSA additive for supporting depolymerization and solubilisation of 2nd and 3rd generation was impressively underlined by comparative oxidation experiments with pomace and cyanobacter. From all results of this study we anticipate a high potential of POM-catalyzed biomass oxidation for future decentralized biomass conversion processes of wet and contaminated substrates. As formic acid is a chemical product used in high amounts by farmers (e.g. in the preparation of silage) and waste streams from farming or food manufacturing represent very attractive feedstocks, the technology appears particularly attractive if linked to agricultural activities or small to medium sized food industries.

Economic and Environmental Assessment of Methane to Ethylene Via Oxidative Coupling

Abstract ID# 469784

Kenneth O'Neill and Benjamin J. Davis, The Cooper Union (New York, NY)

One possible separation train for the purification of ethylene from reactor effluent is proposed, with interest in meeting economic and environmental (via LCA) performance targets. The system consists of two flash tanks and four distillation columns which serve to purify polymer-grade ethylene and reactor recycle. The reaction for producing ethylene is the oxidative coupling of methane on nanowire catalysts, and low conversions mean a large percentage of the methane enters the separation train and must be separated. The four columns following the two initial flash tanks are a carbon monoxide removal column for the separation of carbon monoxide from methane for reactor recycle, a demethanizer column for separation of the remaining methane from the heavier hydrocarbons, a C_2-splitter for the separation of ethylene from ethane, and a carbon dioxide removal tower which operates at vacuum to break the azeotrope which forms between ethylene and carbon dioxide to obtain polymer grade ethylene. The purified ethylene distillate from the carbon dioxide removal tower is 99.481 % pure and the separation system has purchase and bare module costs of $776,220 and $3,300,000, respectively (based on 2015 CEPCI values).

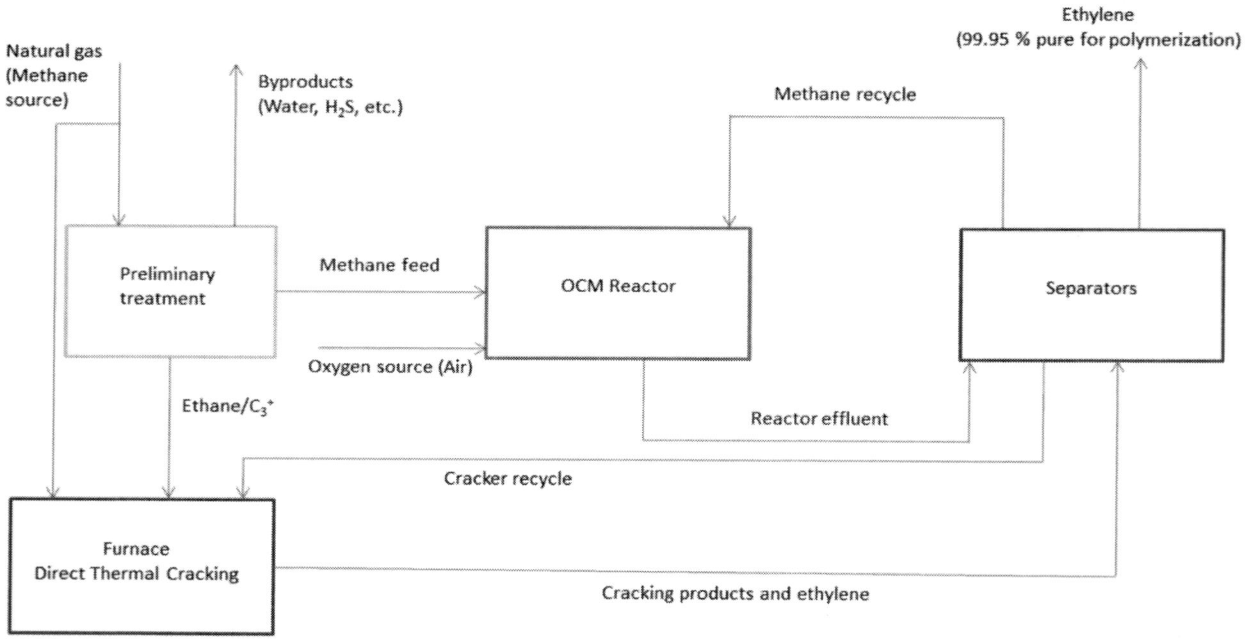

Figure 1: BFD for Siluria's OCM process integrated into an existing ethane cracker

Approximately 90,000 tons of ethylene were produced in United States (US) gas plants in 2014[i], making it one of the world's largest volume commodity chemicals. Ethylene is valuable as a precursor to a variety of products, including low and high density polyethylene, vinyl chloride monomer (VCM), ethylene glycol, and styrene; more than 30 million tons of these commodities were sold in 2012[ii]. Because of its high volume, finding cheaper feedstocks is one of the few ways to have an impact on the ethylene production market. Traditionally ethylene is made from cracking ethane or naptha, which are less abundant species in natural gas and crude oil, respectively. As of February 16, 2016, the sales price of natural gas was $1.95 per MMBTU,[iii] while the total sales of ethylene based polymers decreased 4 % to $69.4 billion in 2014[iv].

Due to the importance of ethylene in the chemical industry, means of revolutionizing and optimizing its production are constantly sought after. Siluria Technologies has created and patented nanowire catalysts[v] and processes[vi] related to the oxidative coupling of methane (OCM) to make ethylene; this is a more advantageous route because methane is cheaper and more abundant than naptha and ethane. A simplified flowsheet from their website is shown below as Figure 2.

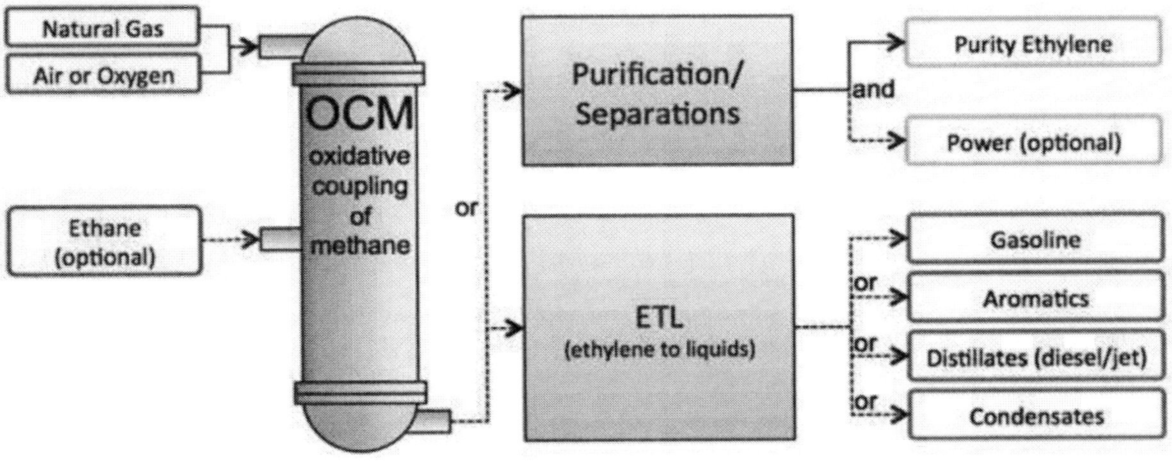

Figure 2: Simplified flowsheet of Siluria's OCM process[vii]

In the Siluria process, the feed to the OCM reactor consists of a methane source (natural gas) and an oxygen source (air or O_2 in the Siluria process). The reactor effluent can then be routed to one of two steps: a separation train for the recovery of high-purity ethylene, or to an ethylene to liquids process to produce gasoline/aromatics/distillates/condensates. The present investigation attempts to propose a more efficient means of separating the reactor effluent for the acquisition of high-purity ethylene.

Traditionally, methane conversions in OCM reactors with ordinary catalysts have been found to vary between 1 % and a theoretical upper bound around 40 % according to Bhasin *et al.* and Lunsford *et al.*, with the remaining 60 % of methane either unreacted, burned, or coked. Selectivities for ethane, ethylene, and acetylene (C_2) tend to inversely with methane conversion, from highs of 90 % to some reports of as low as 1 % have been reported[viii,ix]. At 799 °C, one catalyst patented by Siluria exhibits methane conversión around 21 % with a 70 % selectivity for C_2; on a similar catalyst over an SiO_2 support, the conversion is about 28 % with a 60 % selectivity for at 800 °C. In this study the properties of the former catalyst were used.

With the goal of having an industrially viable OCM system, Siluria has one of two options: construct a stand-alone facility focused on OCM-based ethylene production or integrate the process into a preexisting ethylene cracking facility. With integration into an existing ethylene cracker in mind, the preliminary design investigated herein was one in which the OCM reactor produces about 40 metric tons/day of product (40,000 kg/day).This corresponds to overall plant production of 740 metric tons/day, with 5.4 % of the cracking facility's ethylene from OCM and the remainder from the cracker.
(700 metric tons/day from the cracker).

The primary goal of the research thus far has been to separate the OCM reactor effluent. The difficulty in the separation of high-purity ethylene comes from the complicated OCM reactor effluent mixture. This mixture was assumed to consist of the following species:

Species and Boiling Points	
Component	**Normal Boiling Point (K)**
N_2	77.35
CO	81.15
O_2	90.15
CH_4	111.5
C_2H_4	169.15
C_2H_6	184.15
C_2H_2	189.15
CO_2	194.15
C_3H_6	225.15

Table 1: Components in OCM reactor effluent in order of increasing boiling points

The following assumptions were used in simplifiying the number of components in the reactor effluent:

- Air served as the oxygen source
- Hydrogen was considered light enough to be separated relatively easily
- Water's distinct chemical differences from the hydrocarbon mixture allowed for the assumption that it too could be easily removed
- Propylene was used as a proxy for hydrocarbons of chain length 3 or more, e.g. propanes, butanes, etc., due to those species' relatively higher boiling points allowing for relatively easy separations

In addition to the similarities in properties between the hydrocarbon components, these compounds are also soluble in one another, making separations via distillation more difficult than a simple boiling point difference, making determining appropriate operating conditions for the required separation units difficult. These four simplifying assumptions, along with the results of a material balance around the reactor given below, served as the basis for the simulation.

Component	Mass Flow Rate (MT/day)	Mass %	Molar Flow Rate (kmol/day)	Mole % (OCM Effluent)
CH_4	205.4	42	12800	57
O_2	60.1	0.12	18.8	.083
N_2	226	46	8070	36
CO	3.97	8.2E-01	142	0.63
CO_2	12.3	2.5	279	1.2
C_2H_6	28.1	5.8	938	4.2
C_2H_4	10.079	2.1	360	1.6
C_2H_2	0.041	8.4E-03	1.58	0.0070
C_3H_6	0.082	1.7E-02	1.95	0.0086
Total	487	100	22600	100

Table 2: Mass and molar flow rates and percents for the feed to the separation process simulated via ASPEN Plus V8.0

Figure 3 shows the current separation process as simulated in ASPEN Plus V8.0. The Peng-Robinson equation (P-R EOS) of state was used for the simulation because it works well for light gases[x].

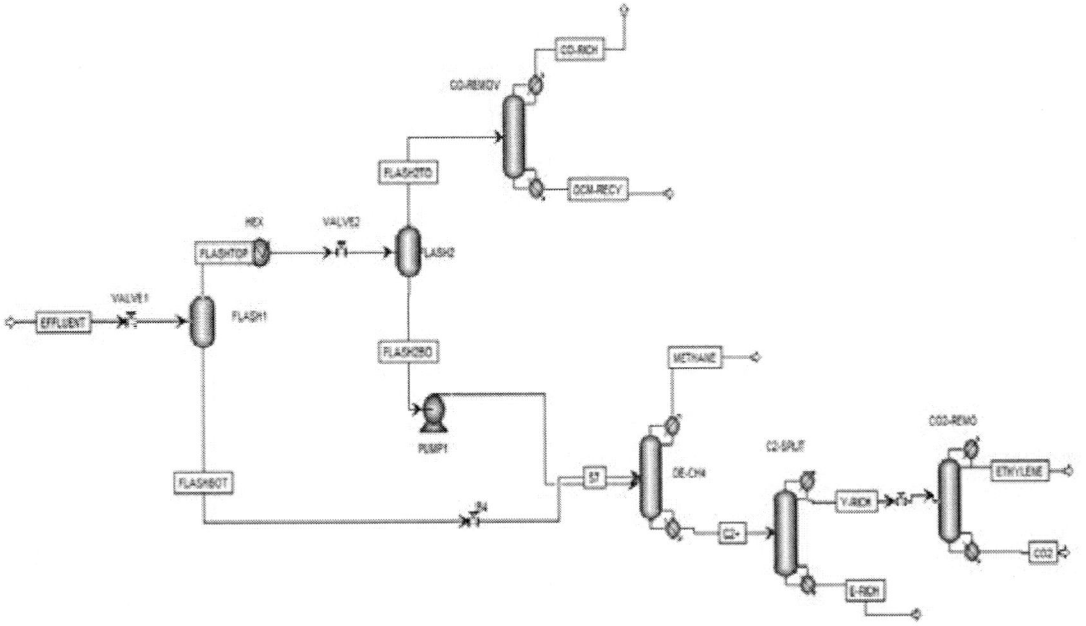

Figure 3: Separation process flowsheet from ASPEN Plus V8.0

The rationale for the flowsheet above was to aim for a good initial separation of light and heavy components, where the ideal case has all of the methane and carbon monoxide leaving as vapor with oxygen and nitrogen, and the remaining components leaving as liquid bottoms from flasg 1. The vapor overhead from the initial flash is cooled, throttled, and flashed again at atmospheric pressure to condense the remaining ethylene; the vapor product is sent to a cryogenic distillation column for the separation of carbon monoxide from the methane and oxygen to be sent as reactor recycle (above the higher flammability limit of methane in air).

The specifications on the individual units are as follows in Table 3 and Table 4.

	Carbon Monoxide Removal	Demethanizer	C2-Splitter	Carbon Dioxide Removal
# of Theoretical Stages	30	10	50	30
Stage Efficiency	0.79	0.77	0.79	0.9
Actual Stages	38	13	63	32
Feed Stage (from top)	17	8, 8	15	24
Reflux Ratio	9	3	3.5	11
Distillate:Feed Ratio	0.14	0.505	0.4	0.544
Condenser P (atm)	2	9.5	9.5	0.08
Condenser T (K)	84	148	220	135
Reboiler P (atm)	2.18	9.555	9.8	0.11
Reboiler T (K)	113	227	241	143

Table 3: Distillation column specifications

Unit	Flash 1	Flash 2	Valve 1	Valve 2	Pump
P or ΔP (atm)	10	2.1	10	1	9.75
T (K)	165	120	-	-	-

Table 4: Flash tank, valve, and pump specifications

From the heavy components (20 % of the feed), CO_2 would preferably be removed as early as possible due to its abundance in the mixture, but due to its high solubility in hydrocarbons and its azeotrope with ethylene, vacuum distillation was necessary to separate it from the ethylene product. The bottoms of the two flash tanks are fed into the demethanizer column to separate the remaining methane (the bottoms of the second flash tank would be pumped back up to 9.75 atm). The distillate from the demethanizer is 99 % pure methane, with the remainder oxygen and nitrogen (above the higher flammability limit) and can be recycled back to the reactor. The bottoms of the tower containing all C_2^+ species would then be sent to a C_2-tower for the separation of ethylene from ethane. Finally, the ethylene and carbon dioxide rich distillate from the C_2-tower would be sent to a vacuum distillation column to break the azeotrope and get a 99.95 % pure ethylene product to be used for polymerization to various grades of polyethylene.

The ethylene separated in the CO_2-removal tower was 99.481 % pure, close to the required purity limit, and would need to be sent through an adsorber prior to polymerization. The CO-Removal tower is processing 90 % of the process flow at cryogenic temperatures and will most likely be the most expensive unit in the system with the consideration of utilties. In addition, the vacuum distillation column for breaking the azeotrope between ethylene and CO_2 will require expensive auxiliary equipment to generate vacuum in the column. The CO-free methane recovered from the bottoms of the CO-Removal tower and the remaining methane recovered as overhead from the demethanizer can be recycled to the OCM Reactor to decrease feed costs. Ethane, acetylene, and other heavier products can be used as feedstock for a traditional cracking furnace to further increase ethylene yields. The current focus of the study is to optimize the designs of the distillation columns in terms of utilities required for the heat exchangers and the power requirements for the vacuum pump.

The columns were sized with 14 inches between trays, 3 feet at the top of the column for vapor disengagement, and 10 feet at the bottom for liquid holdup and structural support. The reflux drums were sized in the same way as the flash tanks previously and the pumps were priced based on the operating pressures and necessary powers.

Distillation Columns				
Specification	CO-Removal	Demethanizer	C2-Splitter	CO2-Removal
Theoretical Trays	30	10	50	25
Tray Efficiency	0.8	0.8	0.8	0.8
Actual Trays	38	13	63	32
R (mole ratio)	9	3.5	3.5	11
Diameter (ft)	6	2.5	3	4.3
Height (ft)	70	32.5	107.5	61
Condenser Duty (MMBTU/h)	-14.3	-1.7	-1.4	-1.5
Reboiler Duty (MMBTU/h)	10.2	2.6	1.4	1.5
Cond. Utility (K)	83.7	147.9	220.4	135.1
Reboil. Utility (K)	112.7	226.9	240.9	142.6
P condenser (atm)	2	9.5	9.5	0.08
P reboiler (atm)	2.18	9.55	9.8	0.11
Purchase Cost	$154,000	$24,100	$243,000	$83,000
Bare Module Cost	$334,000	$58,800	$701,000	$177,000
Reflux Drums				
L/D	5	5	5	5
Volume (ft^3)	40.33	13.87	9.64	4.52
D (ft)	5.71	1.96	1.36	0.64
L (ft)	28.53	9.81	6.82	3.20
Thickness (in)	0.07	0.08	0.05	0.00
Purchase Cost	$23,100	$4,910	$3,390	$2,650
Bare Module Cost	69500	15300	10200	7960

Pumps		
	Demethanizer Pump	Vacuum Pump
Purchase Cost	$6,970	$6,900
Bare Module Cost	$27,800	$20,700

Table 5: Sizing and costing of the distillation columns, their corresponding reflux drums, and the two pumps required in the flowsheet

Finally, after a possible optimal design is made, a pinch analysis for heat integration will be performed to determine the number heat exchangers and amount of utilities required, a full cost-analysis, and LCA will be done. When this is complete, a comparison as a stand-alone plant versus a supplementary ethylene production facility to a traditional ethane cracking plant. Alternative means of separating the CO_2—including membrane separation and the deposition of solid CO_2—will be investigated.

References

[i] U.S. Energy Information Administration. "U.S. Gas Plant Production of Ethane-Ethylene." https://www.eia.gov/dnav/pet/hist/LeafHandler.ashx?n=pet&s=metfpus1&f=a. Feb. 2, 2016. Web.

[ii] Esposito, Frank. "American Chemistry Council reports 2012 resin statistics." http://www.plasticsnews.com/article/20130220/NEWS/130229994/american-chemistry-council-reports-2012-resin-statistics. Feb. 2, 2016. Web.

[iii] U.S. Energy Information Administration. "Natural Gas." http://www.eia.gov/naturalgas/. Feb. 14, 2016. Web.

[iv] UN Comtrade Database. "Polymers of ethylene, in primary forms." http://comtrade.un.org/.Feb. 8, 2016. Web.

[v] United States Patent Application Publication, Scher*et al*. Publication No.: US 2012/0041246 A1, Feb. 16, 2012.

[vi] United States Patent Application Publication, Iyer*et al*. Publication No.: US 2014/0012053 A1, Jan. 9, 2014.

[vii]http://siluria.com/Technology/Overview

[viii]Bhasin, M. M., and Keller, G. E. "Synthesis of Ethylene via Oxidative Coupling of Methane I. Determination of Active Catalysts."*Journal of Catalysis* 73.1 (1982): 9-19. Print.

[ix] Clearfield, Abraham, Hinson, Paul G., and Lunsford, Jack H. "The Oxidative Coupling of Methane on Chlorinated Lithium-doped Magnesium Oxide."*Journal of the Chemical Society*. 20.1 (1991): 1430-1432. Print.

[x] Henley, Ernest J., Junior D. Seader, and D. Keith Roper. *Separation Process Principles*. 3rd ed. Hoboken, NJ: Wiley, 2011. Print.

472330 Highly Efficient Formate Production By Hydrogenation of Captured CO_2

Abstract:

Climate change resulting from the emission of CO2 has become a widespread concern in the recent years. Carbon Capture and Storage (CCS) is a promising technology that has the potential to address the 40% of emission emanating from large-point sources such as power plants. The utilization of CO2 as a raw material in the synthesis of chemicals and liquid energy carriers offers a direct way to mitigate the increasing CO2 build-up but catalytic reduction of CO2 to fuels or value added chemicals is still one of the greatest challenges. Recently, increasing efforts have been devoted to the transformation of CO2 into value added substances such as formic acid or formates, which are highly attractive as commodity chemicals used as preservative and antibacterial agents, as well as clean sources for H2 generation. Herein, an effective low-temperature process of hydrogenating liquid ammonia or amine captured CO2 over supported palladium nano-catalysts, which can leverage the existing industrial infrastructure to produce value-added formate by utilizing the CO2 derivatives in the commercial CCS process, has been developed. For the first time, we found that the yield of formates is highly dependent on the ethyl carbonate ion intermediates in a solution of ammonium carbonate or amine-CO2 adduct. We further designed the rechargeable â€œhydrogen batteryâ€ based on the redox equilibrium between liquid ammonia captured CO2 and the corresponding ammonium formates, where hydrogen was stored in chemical bonds. The stored hydrogen can be catalytically liberated under mild conditions as the feed to fuel cells for regenerating electricity.

Allothermal gasification of lignite and biomass with integrated syngas cleaning and CO_2 removal

Peter Treiber, Michael Neubert, Jonas Leimert, Juergen Karl
Friedrich-Alexander-University, Erlangen-Nuremberg, Germany

Introduction

For the production of synthetic natural gas (SNG) from lignite or biomass in general the following process steps are required [1].

- Gasification of the solid feedstock (allothermal or autothermal)
- Syngas cleaning from particles, tars, and off-gases
- Methanation
- Optional product gas conditioning for grid injection

For small- and mid-scale, decentralized plants the main challenge is to keep the plant complexity to a minimum while still reaching the requirements for gas grid injection. The Heatpipe Reformer technology is a suitable way for decentralized SNG production [2], [3]. A hydrogen rich syngas with a nearly ideal composition for methanation can be produced in the Heatpipe Reformer using allothermal steam gasification. A syngas cleaning unit with integrated removal of CO_2 would reduce the plant complexity and provide syngas for methanation without elaborate gas cleaning steps.

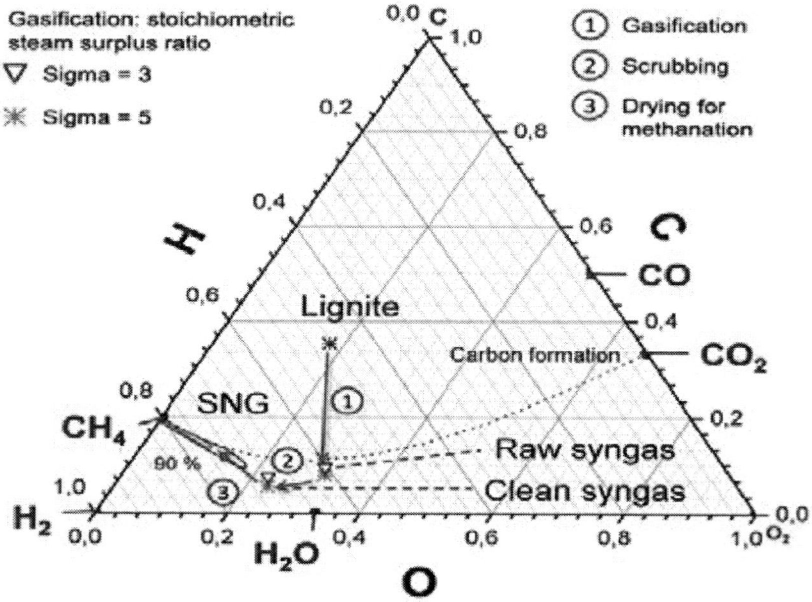

Figure 1: Coal-to-SNG process chain in the ternary diagram

With the EU research project CO2freeSNG2.0 (RFCR-CT-2013-00008), funded by Research Fund for Coal and Steel (RFCS), the research focus of the Chair of Energy Process Engineering at the Friedrich-Alexander-University Erlangen-Nürnberg for the subjects of gasification, syngas cleaning, and methanation is intensified [4]. In addition to the formerly proposed hot gas cleaning [5], [6] a hot carbonate scrubbing unit based on the Benfield process [7], [8] is integrated in the process chain for the combined bulk removal of CO_2, sulfur, and tar species. The proposed process chain and its structure in the ternary diagram are depicted in Figure 1. The tests stands and the results of the parameter variation tests are presented in the following chapters.

5 kW Bench scale process chain

For more flexibility and a reduced test time frame, a process chain for the production of SNG was designed and manufactured in the bench scale for 5 kW fuel input. The schematic setup of the single test rigs and the possible combinations are shown in Figure 2.

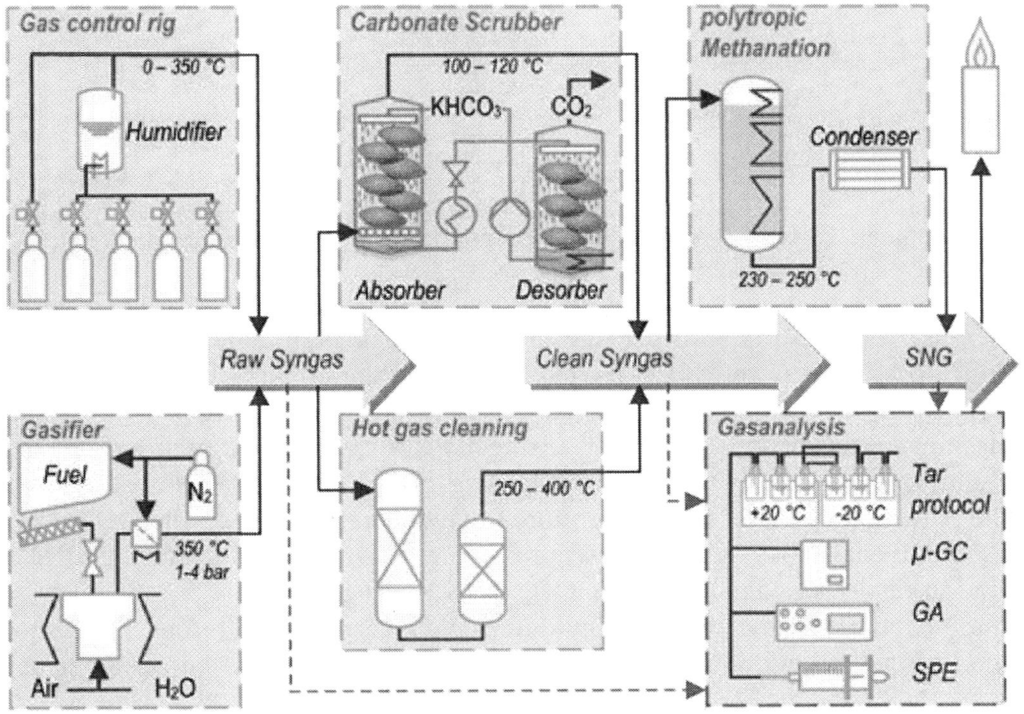

Figure 2: Schematic layout of the EVT 5 kW Coal-to-SNG process chain

Besides real syngas from coal or biomass also bottled gas mixtures can be used in the test rigs. The whole process chain is designed for an operating pressure of 5.2 bar_a. Furthermore, gas sampling ports have been integrated between each of the process steps to analyze gas composition, tar loading and sulfur species in the gas streams.

The basis for the bench scale SNG production is the 5 kW bench scale gasifier. Lignite or biomass can be gasified in a pressurized, steam blown fluidized bed. The regulation of the fluidization behavior as well as operating pressure is integrated in the fully automated test rig which allows for long term tests of over 100 h. The raw syngas from the gasifier passes a hot gas filter where particles and alkali are removed from the gas stream. Subsequently the syngas is passed to the next test rig (scrubber or methanation unit with hot gas cleaning) reducing the pressure between each test stand by 200 mbar to enable an independent pressure regulation for each unit. At the end, the syngas or product gas is disposed over a lab scale flare. The setup and function schematics of the bench scale gasifier are shown in Figure 3.

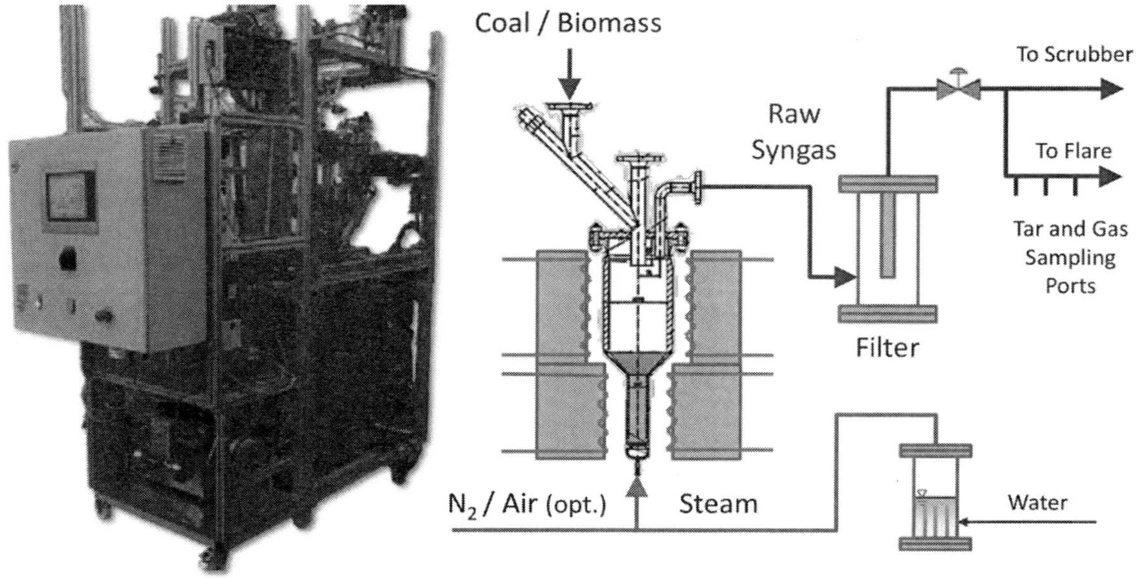

Figure 3: 5 kW bench scale gasifier: setup (left) and function schematics (right)

After the gasifier, the raw syngas is passed on to the bench scale carbonate scrubber (Figure 4) at a pressure of 5.2 bar_a. In the absorber, the raw syngas passes in counter current flow to the solvent over a random packing of 6 mm Raschig rings and exits the column at the top. In contrast to other absorption processes, the solvent is not cooled before entering the absorber (as in amine scrubbing) of the hot potassium carbonate process but it is directly drawn from the reboiler and recycled into the absorber at temperatures only slightly below boiling point (105°C). Therefore, the energy-intensive cooling of the solvent is not needed. Furthermore the regeneration is supported by a partial evaporation of water in the solvent, lowering the partial pressure of CO_2 already at the top inlet of the desorber column. In addition to the absorption of CO_2, also hydrogen sulfide is absorbed and heavier tar components are captured in the solvent flow due to condensation. Therefore, the carbonate scrubbing unit can be considered as bulk removal option for CO_2, H_2S, and heavier tar components.

Table 1: Carbonate scrubbing reactions

$$K_2CO_3 + CO_2 + H_2O \rightleftharpoons 2\,KHCO_3 \quad (1)$$

$$K_2CO_3 + H_2S \rightleftharpoons KHS + KHCO_3 \quad (2)$$

The capture of tar components is mainly based on condensation while the sour gas removal is based on chemisorption with potassium carbonate following the chemical reactions listed in Table 1. The reduced reboiler duty and the omitted solvent cooling are the main advantages of the carbonate scrubbing process compared to other scrubbing processes.

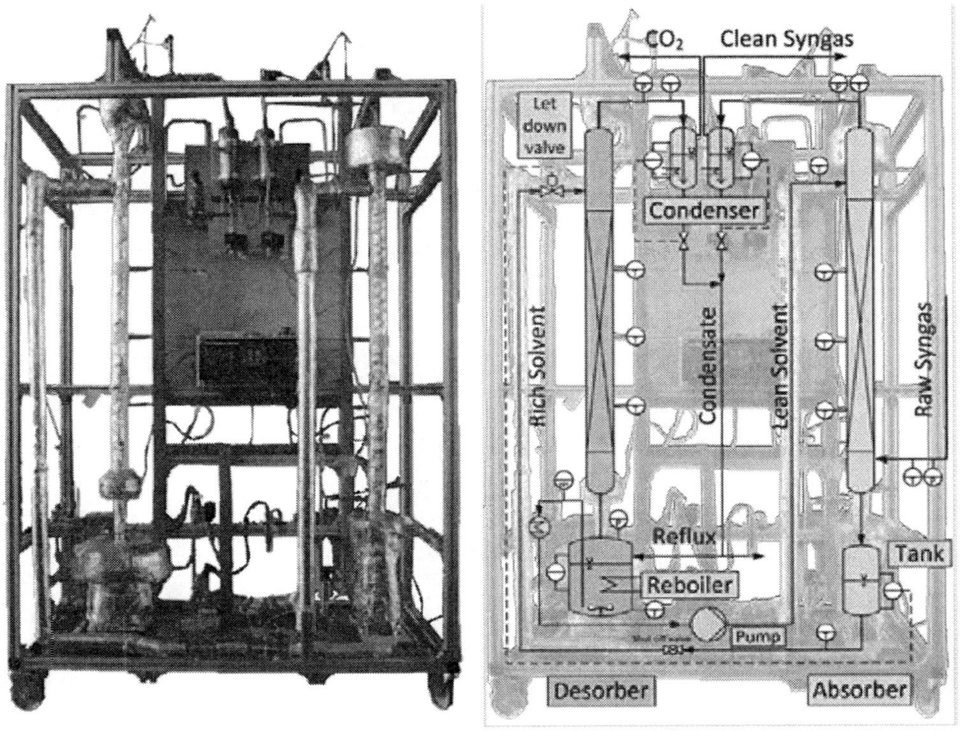

Figure 4: 5 kW bench scale carbonate scrubber: setup (left) and function schematics (right)

100 kW Pre-pilot process chain

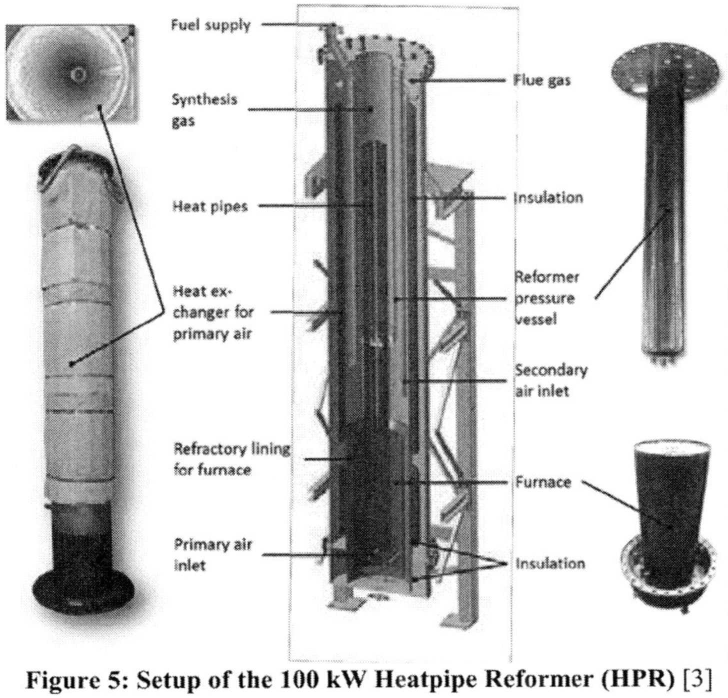

Figure 5: Setup of the 100 kW Heatpipe Reformer (HPR) [3]

Since the 5 kW bench scale test stands are rather prone to heat losses and edge effects (carbonate scrubber columns) the process chain at a 100 kW pre-pilot scale is also investigated. Similar to the bench scale process chain a pressurized, fluidized bed steam gasifier provides the raw syngas. Figure 5 shows the setup of the Heatpipe Reformer. The atmospheric combustion chamber generates the necessary heat which is transported via liquid metal heatpipes to the pressurized gasification reactor at approx. 850°C.

Like in the bench scale process chain, the raw syngas from the Heatpipe Reformer passes over a hot gas filter for the removal of fine coke as well as ash particles. After the filter the syngas enters the absorber column at the bottom inlet. The pre-pilot scale scrubbing unit is shown in Figure 6. The columns (packing height: absorber 3 m and desorber 4 m, diameter: both 100 mm) are packed with 10 mm Raschig rings and have been dimensioned by prior rate based simulations with ASPENplus^{Tm}. The solvent is let down by a throttle valve into a flash drum at the top of the desorber.

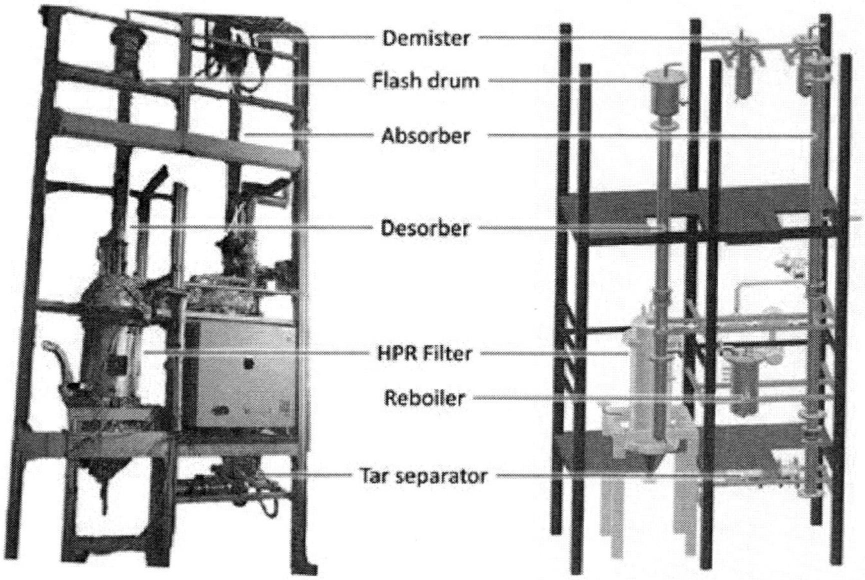

Figure 6: Setup of the 100 kW Benfield scrubber

In contrast to the bench scale scrubber, the reboiler unit of the 100 kW scrubber is not located at the column bottom but is a separate unit which is fed by an additional pump to prevent the enrichment of potassium carbonate in the reboiler. Additionally, a tar separator by means of a coalescence-mesh was integrated in the solvent flow before the main solvent pump. Therefore, a lighter and a heavier tar fraction (compared to the solvent) shall be removed from the solvent flow during regular operation. The total removal of light hydrocarbons is not intended since prior tests have shown that they can be converted in the methanation step [9], [10].

At the top of the absorber column the clean syngas passes a demister and can be either provided for a methanation reactor or disposed in a thermal afterburner. As proposed by Benson and Field [8], [11], [12] a 30 wght.-% potassium carbonate solution shall be used as solvent. Due to reduced solvent volume, the variation of solvent concentration and possible additives is predominantly investigated at the bench scale test rig.

Results

Allothermal steam gasification

In Figure 7 the syngas composition after the gasifier at a stoichiometric steam surplus of $\sigma = 5$ is shown. The relatively high steam surplus causes a syngas which is rich in hydrogen and carbon dioxide.

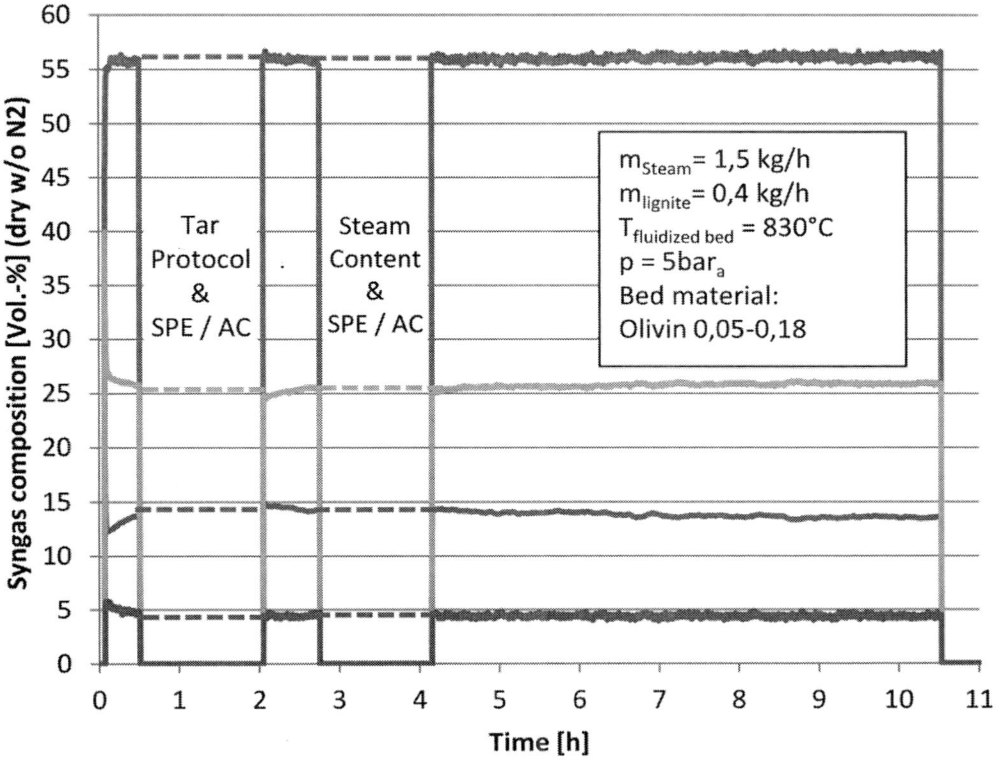

Figure 7: 10-h gasification test at the bench scale gasifier; syngas composition

During the test shown in Figure 7 a steam content measurement and a tar content analysis by means of the tar protocol [13] were performed. The results of the tar protocol measurement are depicted in Figure 8.

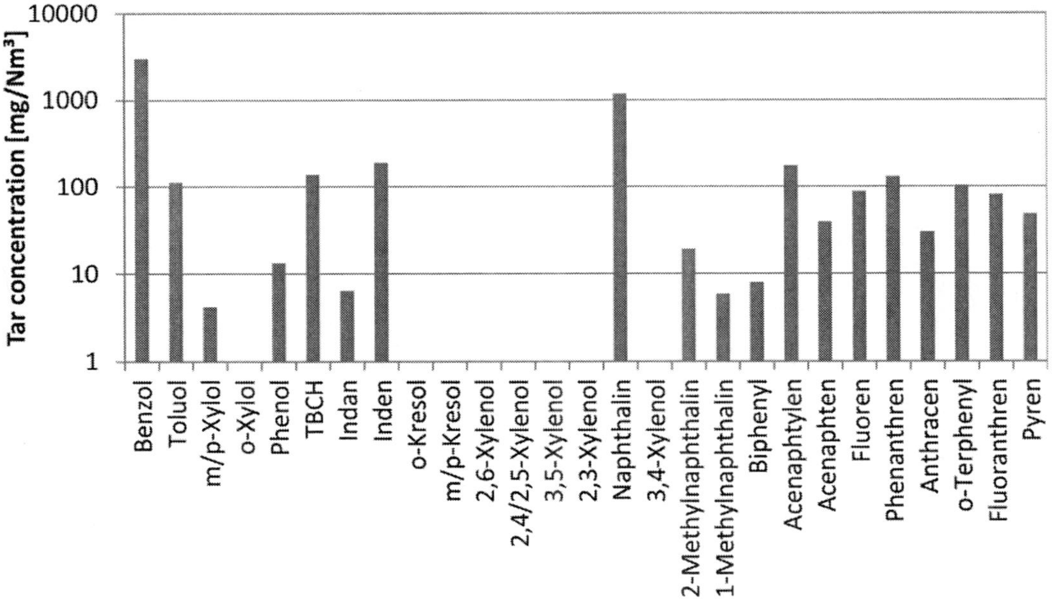

Figure 8: 10 h gasification test, tar concentration in the raw syngas by means of the tar protocol [13]

Integrated syngas cleaning

The influence of the solvent temperature at the absorber inlet as well as the ratio between the solvent and syngas mass flow (liquid-to-gas ratio – L/G) are depicted in Figure 9.

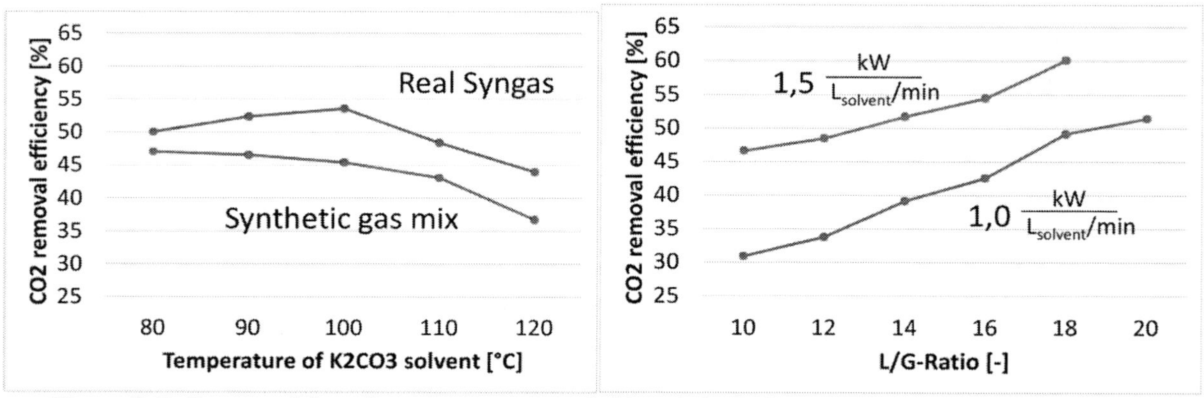

Figure 9: Influence of the solvent temperature (left) and the L/G ratio (right) on the CO2 removal efficiency at the 5 kW bench scale scrubber

At temperatures below 100 °C the CO_2 removal efficiency decreases due to the hindered mass transfer of CO_2 to the solvent and the carbonate reaction rate. At higher temperatures, the solvent loading capacity decreases due to the limited solubility of CO_2.

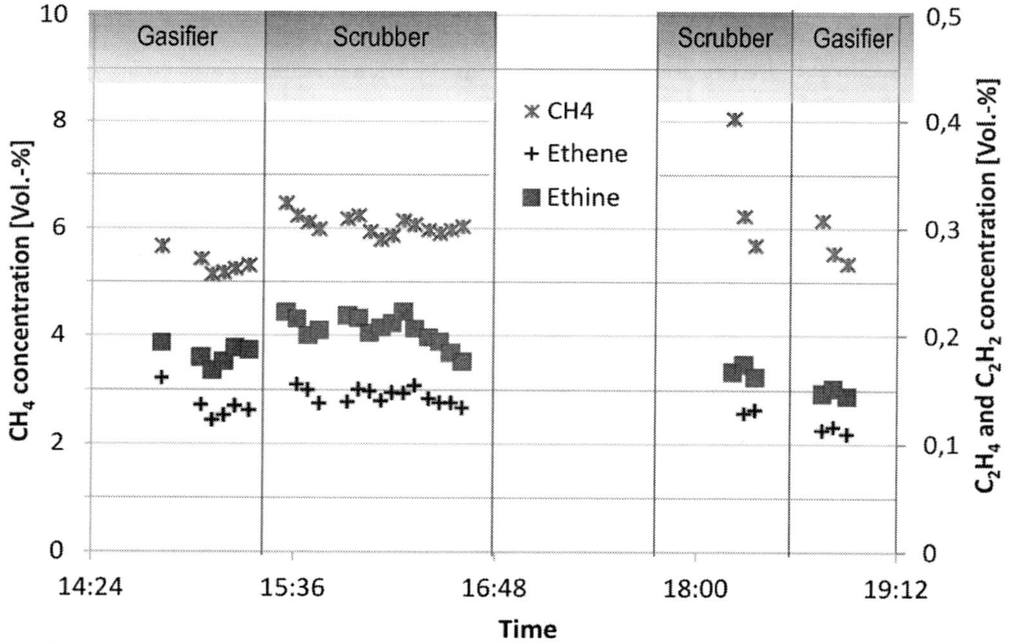

Figure 10: μ-GC measurements after the bench scale gasifier(σ = 5, $T_{gasifier}$ = 830°C, p = 5.2 bar$_a$) and scrubber (L/G = 17, $T_{Solvent}$ = 100°C, p = 5.0 bar$_a$)

Furthermore, a higher removal efficiency of the real syngas could be demonstrated since a slip stream of the raw syngas was used for gas analysis compared to the synthetic tests with bottled gases reducing the syngas volume entering the absorber column. The variation of the solvent flow at constant gas flow proofed an increase in removal efficiency at higher L/G-ratios. An additional increase in the specific reboiler heat duty showed a positive influence on the solvent regeneration and thus an increasing CO_2 removal efficiency.

Compared to the prior simulations, the expected removal efficiency for CO_2 of 90% could not be reached in the bench scale tests. Since the test rig has a relatively small column diameter and the Raschig ring diameter was therefore rather large the scrubber showed dominant edge effects prevented the necessary mass transport in the absorber column. Final tests with the 100 kW pre-pilot test stand are expected to reach the prior simulated removal efficiency.

Besides the direct measurements for the CO2 removal efficiency, also the removal efficiency for other gas components was investigated with a μ-GC. The results shown in Figure 10 demonstrate that light hydrocarbons like ethene and ethine are not influenced by the carbonate scrubbing process and can be measured in constant concentrations in the clean syngas.

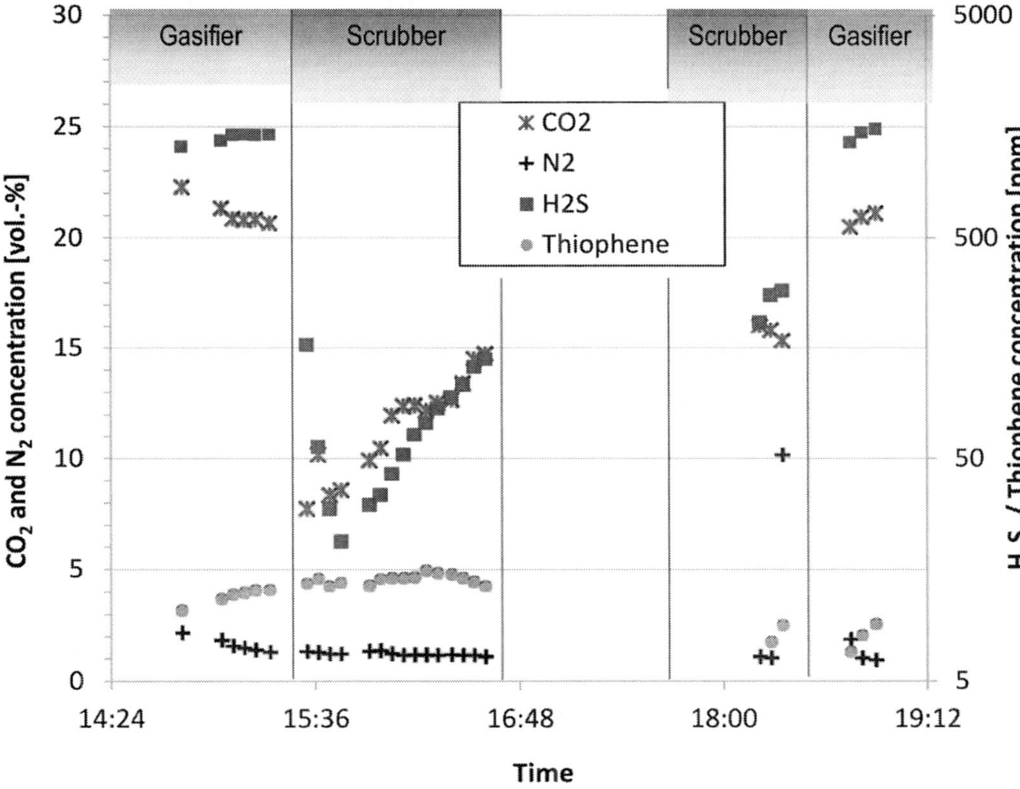

Figure 11: μ-GC measurements after bench scale gasifier and scrubber

The results of the μ-GC measurements in the first half of Figure 11 show the solvent loading process with increasing CO_2 and H_2S content. Since the bench scale scrubber is pressurized and preheated using nitrogen as inert gas, the solvent is stripped to a minimum content of CO_2 and H_2S. In the middle section the product gas after the methanation unit was analyzed showing no content of sulfur species. The final tests demonstrate higher removal efficiencies for hydrogen sulfide than for CO_2 in stationary operation. However the concentration of thiophene was hardly influenced by the carbonate scrubbing unit.

Conclusion

Within the EU research project CO2freeSNG2.0 (RFCR-CT-2013-00008) the process chain for the production of substitute natural gas (SNG) from lignite or biomass for 5 kW und 100 kW fuel input have been constructed. The syngas produced by allothermal steam gasification is treated by means of a hot carbonate scrubbing process and with the same process the bulk share of CO_2 is removed from the syngas stream. First tests could already proof the feasibility of the proposed gas cleaning system which is based on the Benfield process. Besides CO_2 also the bulk share of H_2S was removed from the raw syngas stream while lighter hydrocarbons remained in the gas stream due to higher scrubber temperature. Thus, they are available for conversion in the subsequent methanation step. Organic sulfur components such as thiophene, however, were hardly removed by the carbonate scrubber. Therefore, the proposed process chain demonstrates a bulk syngas cleaning technology for smaller decentralized SNG plants, where conventional syngas cleaning systems are not be suitable due to the high investment costs. In future tests, increased removal efficiency for CO_2 in the 100 kW process chain shall be targeted and the influence of the scrubbing process on tar and sulfur components in the syngas stream will be investigated more thoroughly.

Acknowledgements

The presented test stands and results have been funded by the European Commission and the Research Fund for Coal and Steel (RFCS) as part of the research project CO2freeSNG2.0 (RFCR-CT-2016-00008).

References

[1] J. Karl, *Dezentrale Energiesysteme: Neue Technologien im liberalisierten Energiemarkt*. 2012.

[2] J. Karl, "Biomass heat pipe reformer—design and performance of an indirectly heated steam gasifier," *Biomass Convers. Biorefinery*, vol. 4, no. 1, pp. 1–14, Oct. 2013.

[3] J. M. Leimert, P. Treiber, and J. Karl, "The Heatpipe Reformer with optimized combustor design for enhanced cold gas efficiency," *Fuel Process. Technol.*, vol. 141, pp. 68–73, Jan. 2016.

[4] P. Treiber and J. Karl, "CO2freeSNG2.0," 2013. [Online]. Available: http://www.co2freesng20.eu/. [Accessed: 24-Mar-2014].

[5] C. Baumhakl, J. Karl, T. Kienberger, A. Schweiger, D. Buchholz, S. Bajohr, E. I. Koytsoumpa, and S. Karellas, *Substitute natural gas from coal with internal sequestration of CO 2 (CO2freeSNG) - Final Report*. European Union, 2013.

[6] C. Baumhakl, "Substitute Natural Gas Production with direct Conversion of Higher Hydrocarbons," phdthesis, 2014.

[7] H. E. Benson, J. H. Field, and R. M. Jimeson, "CO2 absorption: Employing hot potassium carbonate solutions," *Chem. Eng. Prog.; (United States)*, vol. 50:7, Jul. 1954.

[8] H. E. Benson and J. H. Field, "Method for separating CO2 and H2S from gas mixtures," US Patent 2,886,405, 1956.

[9] T. Kienberger, C. Zuber, K. Novosel, C. Baumhakl, and J. Karl, "Desulfurization and in situ tar reduction within catalytic methanation of biogenous synthesis gas," *Fuel*, vol. 107, pp. 102–112, May 2013.

[10] C. Baumhakl and J. Karl, "Efficient Bio-SNG Production from Thermal Gasification with Hot-gas-cleaning Technologies," in *21st European Biomass Conference and Exhibition*, 2013.

[11] H. E. Benson and D. H. McCrea, "Removal of acid gases from hot gas mixtures." Google Patents, 10-Jul-1979.

[12] H. E. Benson and J. H. Field, "New data for hot carbonate process," *Pet. Refin.*, vol. 39, no. 4, 1960.

[13] "DIN CEN/TS 15439:2006-08," 2006. [Online]. Available: http://www.din.de/de/mitwirken/normenausschuesse/nmp/din-spec/wdc-beuth:din21:84563381. [Accessed: 01-Apr-2016].

471466 Biocatalytic Polymer Material for Partial Oxidation of Methane to Methanol

Abstract:

An industrial process for the selective partial oxidation of methane under mild conditions would be highly valuable for controlling emissions to the environment and for utilizing vast new sources of natural gas. Methane monooxygenase enzymes perform the difficult activation of methane under nearly ambient conditions, suggesting a biochemical process for methane conversion could be more efficient, less capital intensive, and more easily downscaled compared to current industrial practice. However, the industrial use of enzymes for large-scale applications, especially those involving gas-to-liquid reactions, has been limited by the low enzyme loading and poor mass transfer characteristics of stirred-tank reactors. We have successfully embedded the enzyme particulate methane monooxygenase (pMMO) within a 3D printed hydrogel to create a mechanically robust and gas permeable material that overcomes mass transfer limitations to convert methane to methanol. We studied the kinetics, reactant dependence, and product inhibition of the partial oxidation of methane by pMMO, as these parameters are not well understood and are critical to optimization of our biocatalytic hydrogel material design.

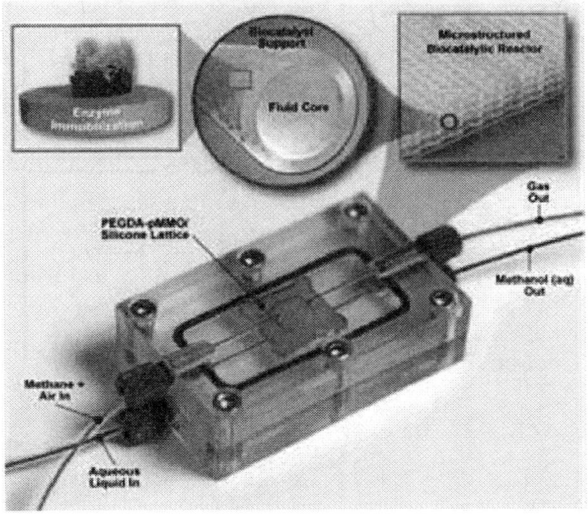

Figure 1: Schematic of enzyme immobilization within hydrogel in a structured polymer scaffold to create a gas-permeable biocatalytic material in a flow-through bioreactor.

Methods: pMMO was immobilized in poly(ethylene glycol) (PEG) hydrogel during crosslinking by UV-initiated photopolymerization. The activity of pMMO was measured by a methanol assay at 45°C using NADH as the reducing agent; methane was injected into the reaction headspace and the methanol produced was quantified by GC-MS. Kinetic studies were conducted by stopped flow analysis with varying pMMO, NADH, dissolved oxygen, and dissolved methane concentrations.

Results and Discussion: Immobilization of pMMO within PEG resulted in a biocatalytic material with nearly 100% physiological enzyme activity. By increasing the surface area to volume ratio of the material using 3D printing, we were able to achieve high volumetric methanol productivity of 2 g/L/hour. Additionally, the hydrogel material can be cured within a poly(dimethyl siloxane) scaffold as part of a flow-through reactor, shown in Figure 1, which is capable of continuous methanol production. Finally, we have determined the dependence of methanol production rate on gas phase reagents, which will facilitate optimization of reactor design parameters such as enzyme loading, membrane thickness, gas-to-liquid volume fractions, surface area to volume ratio, and flow rates.

This work was performed under the auspices of the U.S. Department of Energy by Lawrence Livermore National Laboratory under Contract DE-AC52-07NA27344.

473137 XTL Process with Low Overall GHG Emissions

Abstract:

A patent-pending process that can be used for the production of liquid fuels from biomass/coal based syngas (H_2:CO â %o¤1) over iron-based catalysts while completely suppressing the formation of CO_2 in the FTS process will be presented. Based on the most optimistic data available from Sasol, the fraction of carbon in the original coal that results in the formation of liquid hydrocarbons is only 16.9 % using iron-based catalysts. The data from research employing this new technology have shown that this fraction can be improved to 51.1 %. This is achieved by co-feeding CO_2 (produced in the upstream processes) such that the CO_2:syngas ratio in the reactor is above 3.33:1 and conducting the FTS at conditions above the critical pressures and temperatures of CO_2. The liquids selectivity moved towards higher hydrocarbons (C_{10}- $C_{22}$with peak content in the range of C_{13}-C_{15}). The $ScCO_2$ â€“ FTS process also takes advantage of the strong solvent yet tunable power of $ScCO_2$ and allows for pressure tunable fractionation of the products. A final but important addition to the $ScCO_2$ â€"FTs process is the addition of product gas recycle capabilities that renders the following two unique advantages. (A) recycle helps in improving the throughput (reduced reactor size). (B)Advantages of $ScCO_2$FTS are further enhanced by the recycle of HC gases (along with the unreacted feed) since HCs like hexane and pentane become supercritical. The results from the various stages of development of this technology will be discussed.

474331 Simulation of Hydrogen and Acetylene Production By Methane Pyrolysis in Thermal Plasma

Abstract:

With the recent shale gas disposal and increase of biogas production, acetylene production from methane could play an important role in near future. Indeed, gas pyrolysis by thermal plasma can produce competitive hydrogen and acetylene for PVC production. Besides, acetylene could be converted to ethylene and ethylene to synfuel which can be an attractive gas to liquid solution to produce transportable fuel [1]. A closer process has been already massively set-up and used in Germany before and after Second World War to produce synthetic rubber and hydrogen for coal hydrogenation process [2]. Recently, we have done many experiments converting coal or liquid hydrocarbons to acetylene by thermal plasma with satisfied performances [3]. However, acetylene is unstable at medium temperature and can convert to tar and oil in about 2 ms and then, the pyrolysis products should be cooled rapidly. Cold gas/liquid hydrocarbons can be used as quenching media to increase the yield of ethylene, i.e. a desired by-product. In order to improve the process efficiency, a deep study has been carried out based on experimental and numerical simulation with methane used as feedstock.

The experimental results show that the methane conversion through both plasma devices is higher than 90% and leads mainly to acetylene and hydrogen with high yields. Also, few-layers graphene have been obtained in most of the cases. This investigation suggests that such a simple technique is easy to operate and suitable for mass production of acetylene and few-layer GNs in a scalable process. To improve and have a better understanding of our lab experimental set-up, a numerical model has been established. The plasma is considered using a MHD model. The reaction kinetics and carbon nanoparticle formation have been considered in a sophisticated model. A parametric study has been made on plasma power and hydrogen/methane ratio. The plasma parameters and the hydrogen ratio could be adjusted to improve the acetylene formation and the graphite nanosheets formation

474574 Enabling C_1 Chemistry through Atmospheric Pressure Corona Discharge (non-thermal) Plasmas

Abstract:

Efficient implementation of C_1 chemistry has been a major goal in industrial chemistry for over a century – currently, the major methane to liquids processes involves reforming to syngas followed by catalytic CO hydrogenation. Our group has developed novel technology using atmospheric pressure corona discharge (non-thermal) plasmas implemented in microreactors. In this work we have demonstrated C_1 coupling to C_2 mixtures (ethylene and/or ethane and excess hydrogen, depending on the gas stream composition) with energetic efficiencies in the order of 70%, as well as to longer hydrocarbons. This presentation will summarize recent developments from our lab in this area.

473470 Performance of Hematite Oxygen Carrier for Hydrocarbon Conversion

Abstract:

Hematite has been studied extensively as an oxygen carrier for chemical looping combustion. This study investigates a Canadian hematite under reduction and re-oxidation multi-cycle experiments at a variety of conditions including temperature, space velocity, cycle length, and number of cycles. Post-run characterization is presented to identify which property changes may be most strongly linked to change in performance.

467048 Characterization of Ilmenite As Oxygen Carrier during Chemical Looping Combustion and Reforming

Abstract:

<span" roman"="Roman"" new="New"> Fossil fuel usage is considered a major source of CO_2 emissions, the major contributor to greenhouse effect. Carbon capture and storage (CCS) is considered a promising approach to mitigate CO_2 emissions from such industrial facilities. CCS involves separation of carbon dioxide from the flue gas, transportation to a storage location and subsequent long-term storage. Among these steps, CO_2 capture is the major challenge due to issues relating to the technical feasibility and cost. One of these technologies is chemical looping process (CLP). In the CLP process, an oxygen carrier (a metal oxide) is circulated between two fluidized-bed reactors; a fuel reactor where the oxygen carrier is reduced according to equation (1) and (2), and an air reactor where the reduced oxygen carrier is oxidized according to equation (3). The main advantage with CLP, compared to other conversional conversion processes, is that CO_2 is inherently separated from the other flue gas components, and therefore reducing the economic cost of CO_2separation.

<span" roman"="Roman"" new="New">$CH_4 + 4Me_xO_y \rightarrow CO_2 + 2H_2O + 4Me_xO_{y-1}$(1)

<span" roman"="Roman"" new="New">$CH_4 + Me_xO_y \rightarrow CO + 2H_2 + Me_xO_{y-1}$(2)

<span" roman"="Roman"" new="New">$Me_xO_{y-1} + 1/2O_2 \rightarrow Me_xO_y$(3)

<span" roman"="Roman"" new="New"> CanmetENERGY has two on-going research programs on CLP, one is pressurized chemical looping combustion (PCLC) for the production of power and steam, the other one is pressurized chemical looping reforming (PCLR) for hydrogen production. High reactivity and proper flow

behaviour of the oxygen carrier are crucial in developing both PCLC and PCLR technologies. Among all the low-cost materials including ores and industrial residues, ilmenite, a nature-occurring mineral material, has been paid much more attentions for their potential in CLP due to its desired thermodynamical properties resulting in 100 % theoretical combustion efficiency. Crushed ilmenite ore, henceforth referred to as ilmenite, is considered a leading candidate for oxygen carrying in CanmetENERGY's pressurized chemical looping (PCL) processes since it is relatively inexpensive and possesses good physical attributes. However, as an iron-based oxygen carrier, the reactivity of ilmenite is lower than Mn-, Co-, Ni- and Cu-based ones. Lower reactivity will result in large bed inventory (8-10 times that required for the synthetic Fe-based oxygen carriers) and high pressure drop when used in CFB, and unconverted gas fuels in the outflow of fuel reactor. In addition, there is evidence of attrition, sintering and agglomeration of ilmenite in certain CLP conditions, which negatively impact its reactivity and operating behaviour. This study focuses on introducing nonporous (or mesoporous) dopants of CeO_2, ZrO_2, NiO or Al_2O_3into the ilmenite (mainly immobilized on their surface) to improve the reactivity of the ilmenite-based OCs towards methane combustion and reforming. On the other hand, when mesoporous dopants are employed, the agglomeration of ilmenite particles could be eased.

<span" roman"="Roman"" new="New"> A pressurized thermogravimetric analyzer (PTGA) and a pressurized fixed-bed reactor (PFBR) are used for testing the doped ilmenite particles at various temperatures (850-1050°C), gas compositions, pressures, and cycle stages. These particles are then analyzed by X-ray fluorescence (XRF), X-ray diffraction (XRD), and scanning electron microscopy with energy dispersive X-ray spectroscopy (SEM-EDX). The results from the measurements and discussion on the OC performance are outlined in this paper.

469032 The Role of Support in Increasing Agglomeration Resistance of Cu-Based Oxygen Carriers

Abstract:

Chemical looping combustion (CLC) is a promising alternative to the conventional, *i.e.* combustion based, fossil fuel conversion processes. In CLC, a solid oxygen carrier is used to transfer oxygen from air to a carbonaceous fuel. This indirect combustion route allows for effective CO_2 capture since a sequestrable stream of CO_2 is inherently produced without any need of energy intensive CO_2 separation step. From a thermodynamic point of view, CuO is arguably the most promising oxygen carrier candidate for CLC. However, the main challenge associated with the use of CuO for CLC is its structural instability as the Tammann temperatures of Cu (405 °C), Cu_2O (481 °C) and CuO (526 °C) are lower than the typical operating temperatures of the CLC process. Thus, irreversible microstructural changes are expected for CuO-based oxygen carriers during CLC operation. To avoid the thermal sintering and agglomeration of CuO, it is commonly stabilized by a high Tammann temperature ceramic, *e.g.* Al_2O_3, $MgAl_2O_4$, MgO or ZrO_2. However, the role of support in suppressing agglomeration remains unclear to date because a high Tamman temperature of the support does not always ensure a high resistance to agglomeration. For example, the Tammann temperatures of ZrO_2 and $MgAl_2O_4$ are 1214 °C and 929 °C, respectively. Gayán *et al.*[1] identified that compared to $MgAl_2O_4$ (40 wt. %) a higher weight fraction of ZrO_2 (60 wt. %) was required to manufacture agglomeration-resistant oxygen carriers. In this work, we investigate the mechanism through which a support decreases the agglomeration tendency of Cu/CuO. For this, well-defined surfaces comprising thin films of Cu/CuO and a support (Al_2O_3, Y_2O_3 ZrO_2, MgO, SiO_2) were prepared *via* magnetron sputtering. The cyclic redox stability of the synthesized materials was evaluated in a thermo-gravimetric analyser (TGA) at 400 °C or 900 °C using 10 vol. % H_2 in N_2as the fuel and air for re-oxidation. Energy dispersive X-ray (EDX) spectroscopy on the cross-sections of the thin films revealed that Cu

atoms have the tendency to diffuse outward through the films of the support material during the redox reactions. The support that inhibits the outward movement of Cu/CuO is postulated to provide the highest agglomeration resistance.

1. Gayán P, Adánez-Rubio I, Abad A, de Diego LF, García-Labiano F, Adánez J. Fuel. 2012;96:226-38.

462983 Effects of Structural and Surface Promoters on Manganese-Containing Oxides in Cyclic Redox Reactions

Abstract:

Mn-containing oxides are frequently investigated as low cost and environmentally benign oxygen carriers for chemical looping combustion. The current study investigates Mn-based mixed oxides for two distinct redox applications, i.e. chemical looping with oxygen uncoupling (CLOU) for coal combustion and Chemical Looping based Oxidative Dehydrogenation (CL-ODH) for ethylene production from ethane.

In terms of CLOU, perovskite-structured oxygen carriers with a general formula of A_xAâ€™$_{1-x}Mn_yB_{1-y}O_3$ (A/Aâ€™ = Sr, Ca, Ba; B = Fe, Ni) are investigated. Although $CaMnO_3$ has the tendency to lose its oxygen carrying capacity through irreversible transition to spinel and Ruddlesden-Popper phases, small amount of Sr or Fe dopants enhances the redox stability of $CaMnO_3$. Moreover, the initial oxygen release temperature is significantly lowered, enabling facile oxygen donation and more effective coal combustion. DFT calculations were performed to interpret the dopant effect. It was determined that the aforementioned dopants changes the spatial locations, charge and vacancy formation energy for oxygen anions. Satisfactory correlations among vacancy formation energy, energy barrier for vacancy migration, free energy for phase transition, and redox activity and stability of Mn-containing oxygen carriers are observed.

With respect to CL-ODH, a number of Mn-containing model oxides are investigated. In this cyclic redox scheme, a Mn-based redox catalyst is used to for ethane ODH. Since air is used to regenerate the redox catalyst, CL-ODH avoids the air separation plant. Mg_6MnO_8 based model redox catalysts are investigated in the current study. The effect of various promoters are determined along with the corresponding reaction pathway and underlying mechanisms. It was determined that unpromoted Mg_6MnO_8 redox catalysts are highly selective for CO_2 formation, making it a potentially suitable oxygen carrier for chemical looping combustion. In comparison, promoters such as Na and W significantly improve the selectivity and ethylene yield for the ODH reaction. The overall ODH reaction proceeds through parallel gas phase cracking reaction and selective H_2 combustion on the surface of the promoted redox catalysts.

470135 Morphological Evolution of Transition Metal Nanostructures with Oxygen Dissociation and Migration during Chemical Looping Processes

Abstract:

Transition metal are heavily used in chemical looping technologies because of their high oxygen capacity and high thermal reactivity. These oxygen activities result in the oxide formation and oxygen vacancy formation that affect the nanoscale crystal phases and morphologies within these materials and their subsequent bulk chemical behaviour. In this study, two selected earlier transition metals manganese and cobalt, as well as two selected later transition metals copper and nickel, are investigated with their undergoing cyclic redox reactions. We found Co microparticles exhibited increased CoO impurity when oxidized to Co_3O_4 upon cyclic oxidation; CuO redox cycles prefer to be limited to a reduced form of Cu_2O and an oxidized form of CuO; Mn microparticles were oxidized to a mixed phases of MnO and Mn_3O_4, which causes delamination during oxidation. For Ni microparticles, a dense surface were observed during the redox reaction. The atomistic thermodynamics methods and density functional theory (DFT) calculations are carried out to investigate the effect of oxygen dissociation and migration on the morphological evolution of nanostructures during the redox processes. Our results indicate that the earlier transition metals (Mn and Co) tend to have stronger interaction with O_2 than the later transition metals (Ni and Cu). Also, our modified BrÃ¸nstedâ€"Evansâ€"Polanyi (BEP) relationship for reaction energies and total reaction barriers reveals that reactions of earlier transition metals are more exergonic and have lower oxygen dissociation barriers than those of later transition metals. In addition, it was found that for these transition metal oxides the oxygen vacancy formation energies increase with the depth. The oxide in the higher oxidation state of transition metal has lower vacancy formation energy, which can facilitate to form defective nanostructures. The fundamental understanding of these metal oxide reactions is essential in directing the development of metal oxide-based oxygen carriers for chemical looping applications.

468911 Pt Modified Fe_2O_3@CeO_2 Core/Shell Structures for Effective Thermochemical Water Splitting

Abstract:

Pt modified Fe_2O_3@CeO_2 core/shell structures for effective thermochemical water splitting

Davood Hosseini, Paula Macarena Abdala, Sung Min Kim and Christoph R. Müller

Laboratory of Energy Science and Engineering
Department of Mechanical and Process Engineering, ETH Zurich, Switzerland

Currently H_2 is produced on the large scale largely via the steam reforming of natural gas or the gasification of coal. To yield high-purity H_2, energy intensive purification and separation steps are required. Using chemical looping based approaches, these energy intensive separation steps could be avoided potentially [1]. Chemical looping based H_2 production is a two-step process. In the first step a synthesis gas (*i.e.* a mixture of H_2 and CO) reduces an oxygen carrier, typically a transition metal oxide. In the second step, the reduced oxygen carrier is re-oxidized with steam producing, after the condensation of un-reacted steam, a pure stream of H_2.

In this context, Fe_2O_3 is one of the most promising oxygen carriers owing to its favorable thermodynamics, wide availability and for being environmentally benign. However, at high temperatures (~ 700 – 1000°C) pure, *i.e.* unsupported, Fe_2O_3 deactivates after only a few cycles of reduction and oxidation if reduced fully to metallic Fe [2,3]. One approach to improve the redox stability of Fe_2O_3 is to support it on a high Tammann temperature material, *e.g.* Al_2O_3, ZrO_2, $MgAl_2O_4$ or CeO_2 [1]. The synthesis gas used in the reduction process can be produced *via* the dry reforming of methane (DRM) using an appropriate catalyst, *e.g.* Ni or Pt.

In this work, we report the development of a bi-functional material that allows the integration of the dry reforming of methane into the thermochemical water splitting cycle. The material is composed of a Fe_2O_3@CeO_2 core/shell structure, decorated with Pt nanoparticles. In this work the redox characteristics of the newly developed oxygen carriers were determined at 800 °C using both a thermo-gravimetric analyzer and a fixed bed reactor. To elucidate the effect of the addition of Pt nanoparticles on the reduction of iron oxide, in situ XRD studies were performed.

References

[1] Adánez J., Abad A., García-Labiano F., Gayán P., de Diego L.F. (2012), *Progress in Energy and Combustion Science* **38**, 215.
[2] Bohn C.D., Cleeton J.P., Müller C.R., Chuang S.Y., Scott S.A., Dennis J.S. (2010), *Energy Fuels* **24**, 4025.
[3] Bohn C.D., Müller C.R., Cleeton J.P.E., Hayhurst A.N., Davidson J.F., Scott S.A., Dennis J.S. (2008), *Ind. Eng. Chem. Res.* **47(20)**, 7623.

- Abstract Text

457210 Study on the Performance of $Fe_2O_3/MgO/Al_2O_3$ in Chemical-Looping Hydrogen Generation

Abstract:

Fe_2O_3/Al_2O_3, prepared by the impregnation method, was used as the oxygen carrier in Chemical-Looping Hydrogen Generation. The evaluation performance indicated that the oxygen carrier was active that methane was totally oxidized, and the hydrogen yield was notably high at first. However, carbon deposition was high when reacted with methane and it affected the purity of H_2. The addition of MgO to Fe_2O_3/ Al_2O_3 as a promoter was tested and characterized by BET, SEM, TPR, XRD and XPS methods. Fe_2O_3- MgO/ $LaNiO_3$ was evaluated under the same evaluation conditions as Fe_2O_3/Al_2O_3. The evaluation performance indicated that Fe_2O_3- MgO/ $LaNiO_3$ remained low carbon deposition even after 100 cycles. TPR and XPS characterization illustrated that MgO enhanced the dispersity of Fe_2O_3.

473494 Performance of a Chemical Looping Combustion with Oxygen Uncoupling Process Development Unit

Abstract:

Chemical looping combustion (CLC) is a technology to combust solid fuels while intrinsically separating the greenhouse gas CO_2. This eliminates the energy cost associated with gas separation, making CLC an attractive technology for CO_2 separation. CLC involves circulating a metal-based â€œoxygen carrierâ€ between two reactors in which it is oxidized by air and then reduced by a fuel, in essence combusting the fuel. In the variant of CLC known as chemical looping with oxygen uncoupling (CLOU), the oxygen carrier releases gas-phase oxygen as O_2once in the fuel reactors reducing environment, allowing for a gas-solid reaction to occur. Particles are then circulated to the air reactor and reoxidized. The natural mineral ilmenite presents a cost-effective oxygen carrier material with demonstrated strength and cyclability, while copper allows for the desired oxygen uncoupling behavior of the metal oxide during CLOU operation.

The University of Utah has developed a 10 kW_{th} bench scale dual fluidized bed chemical looping combustion system. The goal of the study presented here was to assess system performance and compare conventional chemical looping combustion, known as indirect gasification-CLC or iG-CLC, with CLOU chemical looping. In this study two different fuels, coal and coal char, were processed and the degree of fuel conversion, CO_2 capture efficiency and CO_2 purity were evaluated. Performance of the unit itself is also evaluated by changing variables such as temperature, flow rate, bed inventory, and circulation rate to determine optimal performance.

469774 Recent Operating Experiences from Netl's 50kW$_{th}$ Circulating Chemical Looping Combustion Test Facility

Abstract:

In order to balance future energy needs with the goal of reducing greenhouse gas emissions, sustainable technology options for fossil fuel power generation are needed. Chemical Looping Combustion (CLC) is widely recognized as a potential option to reduce greenhouse gas emissions at a cost that is significantly lower than a conventional power plant with a conventional CO_2 absorption system. Although the general concept has a long history, CLC has numerous technical issues that must be addressed in order to prepare this technology for a commercial debut. Researchers at the Department of Energy's National Energy Technology Laboratory are using a small 50 kW$_{th}$ circulating test facility to address some of these issues. This paper will describe the test facility and recent operating experience using iron-based oxygen carrier materials. Operating experience has indicated very good gas phase mass balances, as well as, some operational issues associated with solids handling and control. This paper will describe some of the advantages and disadvantages associated with operations using various "commodity" oxygen carrier

472261 The Syngas Chemical Looping Processes for Hydrogen and Electricity Co-Generation with Carbon Capture: Pilot Plant Development and Operation

Abstract:

Chemical looping technologies have evolved into a promising alternative for the efficient conversion of carbonaceous fuels to electricity and/or high value chemicals with minimal carbon emissions. With the exponential growth of research and publications in this field, chemical looping has expanded to encompass power and chemical production with in-situ gas separation. The Ohio State University (OSU) has successfully developed the syngas chemical looping (SCL) and coal direct chemical looping (CDCL) processes from laboratory studies to two fully integrated pilot scale demonstration facilities for the conversion of gaseous and solid fuels, respectively, to high purity H_2 and heat with nearly 100% CO_2 capture. The unique counter-current moving bed reactor design allows the OSU SCL process to achieve full fuel conversion to CO_2/H_2O while minimizing the solid circulation rate and has the capability of producing high purity H_2 for further chemical production.

The present paper summarizes key developments of the 250 kW_{th} â€" 3MW_{th} SCL pilot plant constructed at the National Carbon Capture Center (NCCC). An overview of the process design, cold flow model studies, and construction will be provided. Operational challenges overcome in the SCL pilot plant included the development of auxiliary equipment such as a pilot-scale raw gas burner and baghouse. An advanced high temperature, high pressure solid flow measurement device, electrical capacitance volume tomography (ECVT), was also tested for instantaneous solid flow control in this chemical looping system. The SCL pilot plant has been successfully demonstrated to achieve nearly full syngas conversion. Start-up and operational results including solid circulation studies, process heat up, syngas conversion, and solids analysis will be presented and discussed.

462528 CFD Simulations of a Chemical Looping with Oxygen Uncoupling System: Aiding in Reactor Design

Abstract:

Chemical looping with oxygen uncoupling (CLOU) is a carbon-capture technology that utilizes a metal oxide as an oxygen carrier to separate oxygen from air. The metal oxide oxidizes in an air reactor by the air used to fluidize the reactor and is transferred to another reactor where releases gaseous O_2 into a reactor where fuel, such as coal, is fed. The reduced oxides are then transferred back to the air reactor, completing the loop. Previous experimental and simulated work has been performed on systems less than 1 MW. CLOU kinetics have been introduced into a CLOU dual reactor configuration. These simulations have been used to explore the effects of fuel properties, process properties, and combustion kinetics.

From these simulations design changes to the reactor have been explored including decreasing fluidization volumetric flow rate, altering fluidizing gases, inclusion and design of a carbon stripper. The fluidization volumetric flow rate needed to be decreased proportional to the volatile content of the fuel, this is to maintain proper fluidization regime. The fluidizing gases in the air to fuel reactor loopseal needed to maintain high enough oxygen content as to not introduce oxygen uncoupling in the loopseal resulting in decrease in oxygen transferred to the fuel reactor. From simulations a simple carbon stripper was designed to be used in the system.

473130 Mediated Oxycombustion with Integrated Uncoupled Oxygen Supply (MOBIUS)

Abstract:

In order to sustain the global growth experienced currently, the appetite for energy is going to continually increase. To meet this demand, it is projected that combustion of carbonaceous fuels driven electricity production will continue to play a major role. Coupled with this demand for energy is an increase in the public awareness regarding the need to maintain clean water and air. Therefore, to improve the sustainability of such processes, it is necessary to enhance the efficiency of these power plants and to find reliable environmental compliant solutions. Two such solutions are the Oxy-combustion process that uses oxygen diluted with CO_2 rather than air and the Chemical Looping Combustion (CLC) that employs metal oxide/metal redox couples. There are, however, bottlenecks for both processes. In the former, separate oxygen separation units operated by cryogenic separation or membrane separation becomes a necessity while the latter incurs the technical difficulty of hot solids materials handling as well as performance issues such as reduced conversion rates when dealing with solid fuels. In recent years, there has been a push towards direct carbon fuel cells (DCFC) where the electricity is generated directly in a fuel cell by the direct use of carbonaceous fuels. The primary technological hurdles of DCFC lies in the a) identification of a suitable metal/metal oxide anode system to allow the facile transport of the oxygen from the SOFC electrolyte to the carbon and have a sufficiently high open circuit voltage and b) a compatible electrolyte for the chosen material such the rates of oxidation are high without causing deterioration in the electrolyte by processes such as corrosion. We present a combustion system â€" Mediated Oxycombustion with Integrated Uncoupled Oxygen Supply (MOBIUS) that synergistically combines the effective aspects of the three above mentioned technologies along with innovations to overcome the hurdles faced by these technologies. We will discuss the development of the combustion system components. The key component of this design is a trifunctional membrane reactor based on suitably doped ceria that facilitates the oxygen transport. We will present the impact of doping on the oxygen uptake on the O rich side, its transport through the membrane and

uncoupling at the O lean side.

474591 Upgrading Corn Stover Pretreatment Wastes into Biodiesel Via Oleaginous *Rhodococci*

Abstract:

The derivation of advance biofuels and specialty oils from agricultural waste streams via oleaginous bacteria is a rapidly developing biotechnology primed for both bioenergy and biochemical sectors. However, finding viable effluents and promising microbes that produce an array of tangible oils for subsequent targeting is a current challenge. Here, we examine and compare alkaline and alkaline-peroxide pretreatments of corn stover as a feedstock for three strains of wild-type *Rhodococci*. In comparison, *R. opacus* DSM 1069 and PD 630 exhibited aggressive degradation of low molecular weight aromatics and glucose monomers compared to *R. jostii*, resulting in heightened viability and improved yields. Interestingly, all three strains yielded oils that could be implemented as a fungible diesel post-transesterification or utilized as a specialty oil for novel biochemical/bio-pharmaceutical applications. Overall, this work presents a novel method for waste stream optimization for bio-refineries.

474502 Biological Design of Lignin Conversion

Abstract:

The utilization of lignin for fungible fuels and chemicals remains one of the most imminent challenges in modern biorefinery. The fundamental challenge in lignin degradation lies in the highly recalcitrant structure that requests redox (oxidation-reduction) reactions to break down the chemical linkages. First, we addressed the challenges by revealing the chemical and biological mechanisms for synergistic lignin degradation via a bacterial and enzymatic system, which significantly improved lignin consumption, cell growth and lipid yield. Chemical analysis suggested that laccase, *R. opacus* cell and Fenton reaction reagents promoted the degradation of different functional groups, elucidating the chemical basis for the synergistic effects. Second, we further improved lignin depolymerization by establishing a new enzyme-mediator system to enhance the electron transfer in enzymatic lignin depolymerization to achieve more than 35% solubilization of kraft lignin without extreme temperature, pH, and pressure. The study has manifested that an efficient enzyme-mediator system can be exploited to establish a bioprocess to solubilize lignin, cleave lignin linkages, modify the structure, and produce substrates amenable to bioconversion, all of which are vital for improving current processing and conversion methods for lignin based biofuels and biomaterials. Third, based on the aforementioned studies, consolidated lignin conversion was engineered in both *Rhodococcus opacus* and *Pseudomonas putida* using systems biology-guided biodesign. Efficient ligninase secretion system has been established through proteomics-guided optimization of promoter/rbs, secretion peptide, and others. The engineered strains have led to significantly increased cell growth and bioproduct yield on lignin substrates. Overall, the biological and chemical design of lignin conversion has opened new avenues for lignin conversion toward various bioproducts.

459510 Selective Conversion of Guaiacol to Substituted Alkylphenols in Supercritical Ethanol over MoO_3

Abstract:

Recently, transition metal phosphides, nitrides, carbides and oxides have been intensively employed in biomass conversion{Ma, 2015 #41}.[1-4] The earth-abundant MoO_3 is regarded as a representative of promising alternative catalysts.[2] However, the bottleneck for the development of HDO technique over MoO_3 lies in the consumption of hydrogen and high temperature. In this work, the conversion of guaiacol to various substituted alkylphenols is investigated using MoO_3 in supercritical ethanol at a low temperature of 280 °C in inert nitrogen. Furthermore, the preliminary reaction pathways are proposed and the molybdenum active state is explored.

The conversion of guaiacol was carried out in a batch reactor at a temperature, in 260-300 °C, for a desired reaction time, 1-5 h over a MoO_3 catalyst. The product sample was injected neat into a GC-MS and a GC-FID instrument for analysis. Several possible reaction intermediates (phenol, catechol, anisole, 4-ethylphenol and 2-isopropylphenol) were employed as reactants to understand the transformation path of guaiacol. XRD, XPS and Raman spectroscopy were used to characterize the catalyst samples.

Fig. 1 exhibits that substituted phenols such as ethylphenols, iso-propylphenols, butylphenols (tert-, iso-) and tert-amylphenol were obtained as the dominating products with a high combined selectivity of 86% and at a conversion of 99% over MoO_3 under 280 °C for 4 h in supercritical ethanol.

To explore the reaction pathways of guaiacol conversion, control experiments with phenol, anisole and catechol as the starting materials under the same reaction condition were carried out. Catechol gave a complete conversion while phenol and anisole did not. Besides, a similar product distribution compared to guaiacol as the reactant is achieved with catechol. Considering phenol was never found in the guaiacol conversion process, we propose that catechol is formed as the intermediate via demethylation of guaiacol and followed by direct conversion to alkylphenols without the formation of phenol. Subsequently, we employed 4-ethylphenol and 2-isopropylphenol as the reactants and obtained higher alkylphenols as the products. There were no isopropyl, tertiary (iso) butyl or tertiary pentyl derived aliphatic compounds or n-propyl, n-butyl, n-pentyl substituted phenols detected in the liquid and gaseous products of guaiacol conversion. Thus, we speculate that the higher alkylphenols are from consecutive substitution of lower alkylphenols with methyl and ethyl groups produced by ethanol medium, instead of direct substitution with corresponding substituent groups or isomerization with interrelated alkylphenols (Fig. 2).

The Mo (3d) energy region of used MoO_3 is shown in Fig. 3. As can be seen, it contains large amounts of Mo^{5+} (56.4%) with the $3d_{5/2}$ and $3d_{3/2}$ bands located at 230.9 eV and 234.1 eV, respectively. Moreover, molybdenum trioxide (Mo^{6+} state) and molybdenum dioxide (Mo^{4+} state) have been proved to be inactive in control experiments. Therefore, Mo^{5+} was considered to be responsible for the excellent performance of guaiacol conversion.

468094 Enabling Microbial Utilization of Thermally Depolyermized Lignin Monomers

Abstract:

The lignin component of biomass is an attractive source of carbon and energy for the microbial production of fuels and chemicals. This adding of value (*valorization*) to lignin by biological conversion can increase the overall economic viability of biorenewable fuel and chemical production. Thermal depolymerization of biomass through fast pyrolysis shows promise in terms of its ability to recover aromatic monomers from the lignin component of the biomass. Specialized microbes, such as *Pseudomonas putida* KT2440, can utilize many of these aromatic monomers when provided in pure culture [1,2]. However, microbes can be negatively affected by toxic impurities in biomass derived streams, and lignin rich pyrolysis fractions are insoluble in water and therefore inaccessible to microbes in liquid culture medium. Here we addressed these toxicity and insolubility issues by developing and testing an emulsion formula which granted *P. putida* KT2440 access to the phenolic monomers. The optimized emulsion formula contained 18.2 MΩ water, a 70:30 ratio of Tween20® to Span80® surfactant, and the lignin rich pyrolysis fraction. *P. putida* KT2440 was able to use this emulsion formulation as sole carbon source. Additionally, studies with a pure aromatic monomer, p-coumarate, indicated that the emulsion procedure at least partially protects *P. putida* KT2440 from the substrate toxicity of this monomer. To minimize surfactant cost, the surfactant amount was minimized to 7 wt% of the total lignin rich pyrolysis fraction weight. This emulsification method is an important step in enabling the biological valorization of biomass-derived lignin.

Refs: 1. Johnson, C. W.; Beckham, G. T. *Metabolic Engineering* **2015,** 28, 240-247. **2.** Vardon, D. R.; Franden, M. A.; Johnson, C. W.; Karp, E. M.; Guarnieri, M. T.; Linger, J. G.; Salm, M. J.; Strathmann, T. J.; Beckham, G. T. *Energy & Environmental Science* **2015,** 8, (2), 617-628.

472414 Fractionated Lignin Rheology and Structure

Abstract:

Critical technological challenges remain in the development of commercially viable lignin extraction and utilization processes to yield high value commodity chemicals and novel materials such as aromatics, polymers, adhesives, and flavorings. Processing strategies need to be developed for chemical and enzymatic degradation methods, catalytic upgrading, and purification to make it feasible for commercialization. Designing strategies to process multiphase lignin slurries requires understanding the slurry characteristics, especially quantification of the rheological properties. Fractionating heterogeneous lignin structures yield relatively homogeneous sets that offer simpler processing strategies as compared to inherently heterogeneous whole lignin. The objective of this work was to characterize rheological measurements of fractionated lignin and correlate the rheology to fractionated lignin structures. Organosolv and Kraft lignins were fractionated into five fractions using solvents of increasing polarity. Rheology of lignin fractions dissolved in DMSO was characterized with a rheometer programmed to sweep a range of shear rates for non-Newtonian slurries. ^{13}C nuclear magnetic resonance (NMR) analysis was performed to determine changes in structure and functional group contents of each fractional set. Non-Newtonian behavior increased with increasing lignin concentration within any fractionated group. Although structure varied, there were no obvious rheological differences across solvents or lignin types.

473182 Selective Conversion of Biomass-Derived Lignin to Cyclic Hydrocarbons

Abstract:

A potential value-added stream produced in the DMR process from NREL is the â€œnative-likeâ€ and tractable lignin residues isolated from deacetylation and enzymatic hydrolysis. Nobel metal (i.e.Ru, Rh, Pt) and earth abundant metals (i.e. Cu, Fe, Ni, Zn) immobilized on an acidic carbonaceous material act as efficient catalysts in the hydrodeoxygenation of technical lignins under batch conditions. A synergistic interaction between the metal sites and acid groups within the bifunctional and bimetallic catalyst leads to enhanced catalytic activities for the overall transformation of technical lignins to cyclic hydrocarbons as compared to the individual steps catalyzed by the separate catalytic functionalities. In addition, the selectivity of catalysts will also be addressed in our presentation.

455720 Modeling and Optimization of Reverse Osmosis Desalination: An Industrial Case Study122

Abstract:

A fully-coupled three dimensional Computational Fluid Dynamic (CFD) model is developed to investigate hydrodynamics and mass transfer characteristics in a commercial reverse osmosis membrane. The effects of process parameters on pressure drop and concentration polarization are quantified, based on which a simplified and yet reliable one-dimensional system-level model is developed for industrial-scale analysis and optimization. The theoretical work is successfully validated in a local water plant, demonstrating reduced energy consumption and brine disposal cost.

Reference:

Li, M.; Bui, T.; Chao, S. Three-Dimensional CFD Analysis of Hydrodynamics and Concentration Polarization in an Industrial RO Feed Channel, under review.

465304 Integrated Modeling of Agricultural and Industrial Processes for Intensifying Cane Sugar Industries

Abstract:

We developed a process model of cane sugar industries, which consist of agricultural and industrial processes and produce raw sugar, ethanol and some residues from sugarcane. This model enables to simulate mass and energy flow overcoming the gap between agricultural and industrial processes and to analyze the effects of operational and technological options on the performance of the whole system.

In the previous studies analyzing the life cycle of sugarcane-derived ethanol, the potential of its sustainable production has been confirmed. Although various kinds of research on elemental technologies as well as on their integration approach such as process intensification have been devoted to improving production efficiency, it is difficult to achieve the implementation within engineering fields because cane sugar industries can be represented as complex systems. Various decision-makers are involved in this industry; sugarcane is an agricultural product yielded by farmers, while it is processed into raw sugar and/or ethanol in industrial sectors. The industrial process performance have high sensitivities to changes in the components and conditions of sugarcane, while cultivar choices, environmental conditions such as climate and soil type, and farm operations influence on those properties. Industrial processes also have strong influences on agriculture. For example, many farmers have chosen cultivars which contain much sucrose to meet the demands of the industrial processes. In addition, the material transportation from industrial to agricultural processes is also confirmed. The residues from industrial sectors such as filter cake contain abundant fertilizer nutrients such as nitrogen, phosphorus and potassium, and, therefore, it is implemented worldwide that they are returned to farmlands to maintain the soil fertility.

It is challenging to introduce new technology to achieve high-efficiency production into such complex systems where agricultural and industrial processes are interweaved with each other. The whole system should be regarded as the

objective of intensification and the integrated analysis considering the linkage between them is required to fulfill the potential of the introduced technology sufficiently. Nevertheless, elemental technologies have been developed in the respective sectors up to date. Simulation-based analysis which can visualize the phenomena of agricultural and industrial processes simultaneously is effective to understand the linkage between them. In this study, we have tackled the challenges described above with the development of an integrated model and a case study.

In the agricultural part of the developed model, the relationship among chosen cultivars, cropping type, environmental conditions such as climate and soil, and farm operations such as irrigation, soil pH control, and fertilization was modeled based on the results of the existing reports. The yield of sugarcane and its composition are estimated by this agricultural part. The modeled sugarcane is composed of five components: sucrose, reducing sugars, minerals, fiber and water. The process performance is calculated by the industrial part of the model based on the amount and the ratio of these components. Sucrose and reducing sugars can be converted into ethanol, while reducing sugars affect the efficiency of raw sugar crystallization. Fiber is not a raw material for sugar and ethanol but emitted as bagasse. This is used as a fuel for providing the processes with heat and electricity. The surplus bagasse is generated under some conditions and can be used as a biofuel for other facilities.

A case study using the constructed model demonstrates that an introduction of the new industrial technology into the existing system can achieve significant increase in both sugar and ethanol production as well as the reduction of environmental impacts by choosing appropriate cultivar in agricultural sectors. In other words, the concept of the integrated design proposed in this research can intensify cane sugar industries. This concept offers us a wider range of solutions and help us to improve the performance of the system significantly. The results shown by the model can be used to motivate the real stakeholders such as sugarcane farmers, plant managers and operators of sugar mills, and policy makers to modify their decision-makings in order to increase their profitability and to reduce environmental impacts.

467207 Techno Economic Analysis Activities for New Chemical Processes By Technology Readiness Level

Abstract:

Entrepreneurs and startups in the chemical process industries may, at various points in their development, find themselves asking the question, "What should we be modeling?" While careful planning of fundamental R&D activities is important for startups in any industry, many companies in the process industries face the additional challenge of ultimately operating at huge production volumes. Concurrent activities in conceptual process design and production cost estimation (also known as techno-economic analysis) are therefore also important to ensure early positioning for scale-up.

As a tool for communicating a technology's potential to VCs and other funding agencies, however, the depth of the techno-economic analysis must be consistent with the maturity of its core technology. In other words, concepts that have been well developed in the lab with no consideration given to scale-up challenges may be viewed as insufficiently de-risked. Similarly, a highly detailed analysis presented for a brand-new, unproven concept may be dismissed as too fanciful by inspection.

This talk outlines the techno-economic analysis activities that should be performed concurrent to R&D, using the framework of the Technology Readiness Level (TRL), a system adopted by several U.S. government agencies that provide research funding. The TRL framework ranks a technology's progress toward commercialization on a numerical scale. For example, TRLs 1-3 may represent bench-scale activities, TRLs 4-6 pilot scale, and TRLs 7-9 demonstration and commercial scales. Expected R&D activities are defined at each rank, making it simple for startups to self-identify which TRL applies to them. This talk presents an additional set of definitions regarding the expected process engineering activities at each TRL: when to begin conceptual process design (and at what level of detail), grow process engineering capabilities (in-house or consulting), invest in expensive process simulation resources, and ultimately approach an EPC firm to begin plant construction.

473119 Co-Solvent Hydrothermal Liquefaction of a Microbial Yeast Biomass

Abstract:

Direct hydrothermal liquefaction (HTL) is an effective route that converts whole, wet biomass into an energy-dense liquid fuel precursor, called 'biocrude'. HTL represents a promising alternative to conventional lipid extraction methods as it does not require a dry feedstock or additional steps for lipid extraction. HTL is normally performed in hot-compressed (subcritical) water at a 280-370 °C temperature and 5-25 MPa (700-3000 psi) pressure, although the actual operating conditions largely depend on the desired end product and the initial feedstock type and composition. High working pressure can create challenges in reactor sizing and overall operating costs. Through the use of co-solvents the HTL operating pressure can be reduced. The present study reports results from low-temperature co-solvent HTL treatment of oleaginous yeast, *Cryptococcus curvatus*, conducted at 240-300 °C using laboratory batch reactors. Isopropanol-water was used for the co-solvent HTL experiments and Na_2CO_3 was used as a catalyst. Co-solvent HTL produced significantly higher yield of biocrude (56.4±0.1%) than that of HTL performed without a co-solvent (49.1±0.4%). Addition of Na_2CO_3 as a catalyst marginally increased the biocrude yields, by 7.1% in HTL without a co-solvent and 2.7% in HTL with co-solvents. The energy content of the produced biocrude was ~37 MJ kg^{-1}, which is significantly higher than that of the original feedstock, 24 MJ kg^{-1}. Gas chromatography–mass spectrometry of the biocrude suggested that the HTL process was successful in removing carboxyl groups from fatty acids and creating their associated alkanes (eicosane and heptadecane). Experimental results were leveraged to inform techno-economic analysis (TEA) of the baseline HTL conversion pathway to evaluate the commercial feasibility. Results from TEA models show a renewable diesel price of $4.12 per gallon with the HTL processing accounting for approximately 15% of the total cost.

Optimal supply chain in electric power generation from biogas produce from manure in La Ciénega Michoacán

Pascual Eduardo Murillo-Alvarado[1*], Melitón Estrada-Jaramillo[1], Iván Vera-Romero[1]

[1]Energy Engineering, Universidad de La Ciénega del Estado de Michoacán de Ocampo, Sahuayo, México, 59103.

* Corresponding author. *E-mail*: pemurillo@ucienegam.edu.mx

Abstract

La Ciénega region in Michoacán is one of the most important regions for the production of cattle is why through this paper intends to consider the manure of these regions for the production of biogas. In addition, the biogas can be used for energy production through an internal combustion engine. For this reason, a mathematical optimization model is proposed, the mathematical model determine the supply chain considering economic, environmental and social aspects, through the installation of different biogas generation plants, as well as distributed and central power plants for power generation, the energy obtained is sent to areas with demand for electricity.

Introduction

The main environmental problem today is the pollution and the global warming, the carbon dioxide (CO_2) released the atmosphere is one of the main causes of the greenhouse effect and the principal factor of the global warming, this and other toxic compound are released in significant quantities during the burning of fossil fuels, which have been the main source of energy until today. However, it is expected depletion in reserves of fossil resources in the short term. Therefore, it is expected that from biofuels can reduce the negative environmental impact of the use of energy from fossil resources, these biofuels are produced from biomass and species of agricultural use such as corn, soybean, sunflower , palm, etc., because of their high carbohydrate content. Biofuels production and use increased from biomass are bioethanol and biodiesel. Therefore, the use of biomass can be the most promising way for the creation of an industry from biomass. With the best use of biomass in the energy sector it is possible to obtain numerous benefits for the sustainable development of the energy sector. In this sense, some research has presented the use of biomass as a source for the energy sector, have recently been reported several investigations focused on the processing of biomass as feedstock for the production of different products. Huber et al. (2006), Saxena et al. (2008) and Goyal et al. (2008) presented different ways to produce biofuels and biochemicals. Some reports such as Crabtree and El-Halwagi (1994) present several techniques for the synthesis of components from biomass taking into account the environmental impact. Also, Pistikopoulos et al. (1994) and Buxton et al. (1997) presented a methodology for minimizing the environmental impact considering alternatives reaction. It is known that disjunctive programming techniques are important to effectively solve optimization problems. Therefore, in the case of optimization techniques and logical decision, there have been numerous investigations, works such as Raman and Grossmann (1994) and Lee and Grossmann (2000) presented techniques based logic programming modeling. Recently, there have been works in which a systematic analysis is performed to determine the optimal route for a biorefinery taking into account the available raw material, intermediate products and routes of different processes. In this context, Pham and El-Halwagi (2012) presented an optimization methodology to determine the configuration of a biorefinery through a systematic approach. Ponce-Ortega et al. (2012) and Murillo-Alvarado et al. (2013) presented a systematic optimization model to determine the optimal configuration of a biorefinery considering several possibilities to determine the optimum route processing. Some investigations are aimed at the study of the production of biofuels, mainly bioethanol, as it is currently one of the most in-demand biofuels. In this regard, Huang et al. (2009) studied the effect of different species of biomass and the particle size in cellulosic ethanol production.

Currently, most research in the field of biofuels are focused on the production of biogas as an alternative for reducing fossil fuels, which currently has increased demand due to the multiple uses for power generation, heating , etc. Biogas is generally obtained by the biodegradation of organic material, so that one of the main sources as feedstock for biogas generation has proven to be waste and livestock waste, which can generate substantial environmental and economic benefits for the regions cattle. The current research in this area considering the biogas production from organic biomass, Amon et al., (2004) take into account the biogas production from agricultural biomass by the optimization of methane production from maize and the dairy cattle manure, Gelegenis et al., (2007) considering the biogas production and the optimization, the biogas production is from olive-mill wastewater using different conditions of biomass decomposition. In the biogas production is important considering the best operational conditions to obtain a better range of biomass conversion, in this study is proposed the manure transformation in anaerobic conditions.

To obtain the raw material of this case study is considering the region of La Cienega is a region made up of some municipalities in the eastern part of Jalisco and Michoacán in Mexico, this region is characterized by having a great potential in agricultural production, so it is one of the regions that have a significant economic growth. The region of La Cienega in the state of Michoacan includes the cities of Briseñas, Cojumatlán, Jiquilpan, Pajacuarán, Sahuayo, Venustiano Carranza, Villamar, Vista Hermosa and Ixtlán, which represent the cities that provide the raw material for the case study proposed.

Case Study

The case study proposed considering the following aspect, the region of La Cienega de Michoacan has a high production of cattle and pigs in this research considering the use of manure produced by livestock as raw material for the production of biogas, which is produced through anaerobic digestion of the raw material. In addition, the biogas produced can be used for energy production, which turns out to be a renewable source with great potential for energy production. In this sense, biogas production through livestock manure can be high both economic and environmental benefits, this because it can reduce the use of non-renewable energy and a reduction in the distribution of pollutants because manure can be deposited in places not established generating a problem of environmental pollution and health.

This study also propose a mathematical optimization model to determine the optimal supply chain for power generation through biogas produced by waste and livestock waste. The proposed mathematical model considers the economic benefits of energy production and biogas, which is considered the biogas generation costs and the possible installation of plants where biogas production process is carried out. In the case of power generation, the proposed model considers the possible installation of power plants, where it is considered applying different technologies for power generation, therefore the model is able to determine the best alternative for power generation considering greater economic benefit and reduced environmental impact. Finally, the energy obtained from biogas can be sent central and distributed plant form electricity generation; this electricity produce is sent to markets that demand energy. The mathematical model proposed, also considering the possible installation of the central and distributed plants, this plants can produce energy using a internal combustion engine, for these reason the model proposed will be determine the best supply chain considering all the aspects involved to obtain electricity by the manure from the cities of La Ciénega. In addition, the biogas production process considering the use of water, for the reason in the case study is presented as use of fresh resources to assess the environmental impact due to the external use of water as well as cost. Once the biogas production process is done, the remaining waste can be sold as fertilizer which is a supplement to the economic benefit due to the application of this technology.

Optimization Formulation

The model proposed takes into account the Figure 1, and the model considering the following sets: the set *a* represent the areas where is proceedings the raw material, the set *b* represent the plant where is generated the biogas considering the manure of the regions *a*, the set *c* represent the central plants for the energy generation, the set *d* represent the distributed plant for the energy generation, the set *m* represent the markets with energy demand and the set *t* represent the period time:

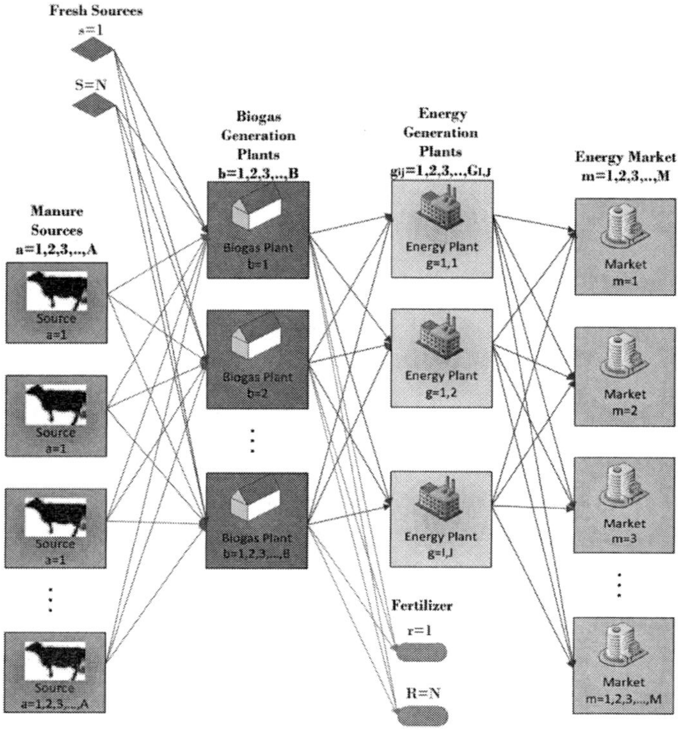

Figure 1. Superstructure for bagasse distribution.

The following equation represents the balance of the raw material (manure) available in the different regions *a* in La Cienega that have a great concentration of manure:

$$FManure_{a,t} \leq \mathrm{FManure}_{a,t}^{MAX}, \forall a \in A, t \in T$$

The mas balance of raw material that is sent from the regions *a* to the biogas generation plant *b* is sets as follows:

$$FManure_{a,t} = \sum_{b \in B} FManureBGP_{a,b,t}, \forall a \in A, t \in T$$

The following mas balance considered the flow of fresh sources f that will be used in the process for the biogas obtaining.

$$FreshSources_{f,t} = \sum_{b \in B} FlowFresch_{b,f,t}, \forall f \in F, t \in T$$

The total mass balance in the biogas production plant considering the flow of the total manure in the regions *a*, and the flow of fresh sources. The total mas balance is set as follows:

$$FTotalManureBGP_{b,t} = \sum_{a \in A} FManureBGP_{a,b,t} + \sum_{f \in F} FlowFresh_{b,j,t}, \forall b \in B, t \in T$$

The following equation calculates the flow of the biogas obtained it the flow of manure in the plants b, is considered the efficiency on the biogas production $\alpha\mathrm{Bio}_{\mathrm{b}}$, the value of αBio_b is calculated according with experimental data:

$$FTotalManureBGP_{b,t} \cdot \alpha\mathrm{Bio}_{\mathrm{b}} = FBiogasBGP_{b,t}, \forall b \in B, t \in T$$

The following expression calculate the flow of compost obtained in the plant, this flow of compost represent de waste of the process to which is not possible to obtain more biogas in the production plants, the total flow of compost $FCompstBGP_{b,t}$ is calculated considering an efficiency in the biogas production in waste $\alpha\mathrm{Compst}_{\mathrm{b}}$:

$$FTotManureBGP_{b,t} \cdot \alpha\mathrm{Compst}_{\mathrm{b}} = FCompstBGP_{b,t}, \forall b \in B, t \in T$$

In the following expression is calculated the total flow of the biogas obtained in the process plants $FBiogasBGP_{b,t}$. Also, this flow of biogas obtained in plants is send to the plants to obtain electrical energy considering distributed and central system:

$$FBiogasBGP_{b,t} = \sum_{c \in C} FBioCent_{b.c,t} + \sum_{d \in D} FBioDist_{b,d,t}, \forall b \in B, t \in T$$

The total mass balance in central plant to obtain energy from the biogas obtained in plants *b* is sets as follows:

$$FTotalCentralEG_{c,t} = \sum_{b \in B} FBioCent_{b.c,t}, \forall c \in C, t \in T$$

The balance considering in the central plant to obtain energy considering efficiency factor $\alpha\mathrm{EGTechC1}_{\mathrm{c}}$, that represent the technology available to obtain energy from biogas in central plants, the balance take into account the following expressions:

$$FTotalCentralEG_{c,t} \cdot \alpha\mathrm{EGTechC1}_{\mathrm{c}} = EnergyCentT1_{c,t}, \forall c \in C, t \in T$$

The following equation represents the total mas balance in the distributed plants to obtain energy from the biogas obtained in plants b:

$$FTotalDistEG_{d,t} = \sum_{d \in D} FBioDist_{b,d,t}, \forall d \in D, t \in T$$

The production of energy from biogas in distributed plants considering the technology available, and the balances is sets as follows:

$$FTotalDistEG_{d,t} \cdot \alpha\mathrm{EGTech}D1_{d} = EnergyDistT1_{d,t}, \forall d \in D, t \in T$$

The total energy obtained in the central energy plants is sending to the markets that present a demand of energy. The following balance represents the energy of central plants that is send to the markets.

$$TotalEnergyCent_{c,t} = \sum_{m \in M} EnergyCent_{c,m,t}, \forall c \in C, t \in T$$

At the same that in previous case, is calculate the balance of the energy that is send to the markets from distributed plant.

$$TotalEnergyDist_{d,t} = \sum_{m \in M} EnergyDist_{d,m,t}, \forall d \in D, t \in T$$

The total balance of energy in the markets, considering the energy from distributed and central plants is sets by the following expression:

$$EnergyMarket_{m,t} = \sum_{c \in C} EnergyCent_{c,m,t} + \sum_{d \in D} EnergyDist_{d,m,t}, \forall d \in D, t \in T$$

The total energy production is according with the demand of energy in the markets, for this reason is proposed the following constraints that represent the maximum and minimum limit in the energy demand.

$$EnergyMarket_{m,t}^{MIN} \leq EnergyMarket_{m,t} \leq EnergyMarket_{m,t}^{MAX}, \forall m \in M, t \in T$$

Finally, is proposed the objective function that considered the total cost and the total sales, the objective function consist in maximize the profit. The total cost is set with the following expression:

$$TotalCost = \left\{ \begin{array}{l} TotalOperCost + TotalCapitalCost + CostTranspMan + CostTranspBiogasCent + \\ CostTranspBiogasDist + CostTranspEnCent + CostTranspEnDist \end{array} \right\}$$

The sales are calculated with the following expression:

$$TotalSales = \left\{ \sum_{b \in B} \sum_{t \in T} FCompstBGP_{b,t} \cdot \beta_{b}^{abono} + \sum_{m \in M} \sum_{t \in T} EnergyMarket_{m,t} \cdot \beta_{m}^{Energy} \right\}$$

$$PROFIT = TotalSales - TotalCost$$

Results

The model is programmed in GAMS through the solver CPLEX, since it is MILP model. The results show that the production of biogas from manure in these cities of La Ciénega can bring great economic and environmental benefits, as it can cover between 5 and 10 percent of total energy demand if these resources were well used. The use of such waste in biogas production can avoid the use of fossil fuels in energy production and thus reduce the environmental impact. In addition, a large percentage of earned profit gained from sales of final process residues, which are sold as fertilizer, which also brings an environmental benefit due to the decrease of toxic components. The electrical energy that can be generated through biogas produced by manure from cattle and pigs can replace approximately 4.23% of the entire annual consumption of the La Ciénega region.

References.

1. Crabtree, E.W.; El-Halwagi, M.M. Synthesis of environmentally acceptable reactions. *AIChE Symposium Series*. 1994, 90 (303), 117-127.
2. Goyal H.B.; Seal D.; Saxena R.C. Bio-fuels from thermochemical conversion of renewable resources: A review. *Renewable and Sustainable Energy Reviews*. 2008, 12 (2), 504-517.
3. Huber G.W.; Iborra S.; Corma A. Synthesis of transportation fuels from biomass: Chemistry, catalysts, and engineering. *Chemical Reviews*. 2006, 106 (9), 4044-4098.
4. Lee, S.; Grossmann, I. E. New alghorithms for nonlinear generalized disjuntive programming . *Computers and Chemical Engineering*, 2000, 24(9-10),2125.
5. Saxena, R.C.; Seal, D.; Kumar, S.; Goyal, H.B. Thermo-chemical routes for hydrogen rich gas from biomass: A review. *Renewable and Sustainable Energy Reviews*. 2008, 12 (3), 1909-1927.
6. Raman, R.; Grossmann, I.E. Modelling and computational techniques for logic based integer programming. *Computers and Chemical Engineering*, 1994, 18 (7), 563-578.

471360 Optimal Design of a Fixed-Bed Ilmenite-Based Chemical-Looping Combustion Process

Abstract:

Global warming is a serious threat to the environment and human life. Inordinate emission of greenhouse gases as a result of increasing global energy demand is a major factor contributing to global warming. Although renewable energy sources are actively being developed, fossil fuels are expected to continue being a predominant part of the global energy mix in the next several decades. Emitted carbon dioxide from fossil fuels combustion is the one of the most influential greenhouse gas which leads to global warming. Chemical-looping combustion (CLC) has been suggested as one of the most efficient methods for carbon capture. CLC is a nonconventional unmixed combustion process where the fuel and the air reactions occur in separate reactors. In this manner, the dilution of combustion products with nitrogen is avoided and the exhaust gas is highly concentrated in CO_2.

Fluidized-bed reactors are commonly used for CLC processes, since the small size of the particles used in fluidized beds provides a good contact between the solid and gas phase. In addition, the rapid mixing of solids ensures adequate temperature control. Fluidized-bed reactors for CLC applications, however, face a number of challenges such as the availability of high temperature and high pressure solids filtering systems that can remove the fines produced from particle attrition. More importantly, using interconnected fluidized-beds at high pressure pose some challenges such as difficulty in finding a stable circulation of solids between the pressurized reactors. At high operative pressure, fixed-bed reactors might be preferable as the attrition of the oxygen carriers will be negligible and hence cyclones and filters downstream will not be required. Heat management and temperature control in fixed-bed systems, however, can be challenging since oxygen carriers are stationary with no mixing and hence bed temperature will be varying, which might lead to process instability.

Nickel and cooper based synthetic minerals (e.g. NiO and Cu_2O supported in alumina) are commonly chosen as oxygen carriers due to their high reactivity with methane and high oxygen carrying capacity (0.21 kg O_2/kg of Ni carrier). However, the cost of these synthetic carriers can be very high (around $19/kg for Ni carriers), limiting the economic attractiveness of the CLC process. Ilmenite is a natural mineral commonly found in igneous rocks and composed of iron and titanium oxide ($FeTiO_3$). Due to its relative natural abundance, ilmenite has a lower cost than synthetic carriers at around $0.3/kg. Furthermore, it has a low tendency towards carbon deposition, increasing the number of cycles that the carrier can be used without regeneration. It also has a relatively high oxygen carrying capacity at around 0.05 kg O_2/kg of carrier.

High pressure chemical-looping combustion of methane using ilmenite as oxygen carrier was investigated in a set of five dynamically operated fixed-bed reactors. Each reactor cycles through four operating stages: reduction, reforming, oxidation, and heat removal. The steam reforming stage allows the bed to be cooled down before the next oxidation stage. Reactor geometry (L/D ratio), catalyst load, and particle size diameter were selected as model parameters; input operating conditions of 10 kg/s of methane at 350°C and 20 bar were assumed. To maintain the long-term chemical and mechanical stability of the ilmenite oxygen carrier, gas temperature along the reactor was kept below 1200°C.

A dynamic model accounting for the change in the area of the intraparticle reaction front as reaction proceeds, intraparticle diffusion, heat transfer limitations inside the oxygen carrier, and spatial variations along the reaction bed was developed. The dynamic model was solved numerically, and the temperature and compositions profiles were obtained at each stage of the CLC process. Sensitivity analysis showed that by adjusting flue gas recycle rates the temperature profiles and heat evolution can be managed in an optimal way. It was concluded that during the oxidation stage, which is highly exothermic, the relevance of heat transfer limitations is negligible but internal diffusion effects should be accounted for. In the reduction stage, where high CO_2capture efficiency is sought, it is necessary to take intraparticle mass transfer limitations into account. The use of ilmenite -a low cost, naturally occurring, oxygen carrier- has a significant potential to improve the performance and feasibility of chemical-looping combustion systems.

462654 A Purge Bin Model for the Recovery of Polyolefins

Abstract:

In the manufacturing of polyolefins, the product purge bin is a challenging unit operation to model because of the counter-current mass transfer at low concentrations of residual components in the gas/solid system.

After the reactions of monomer (and co-monomer) over a catalyst in the presence of inerts (including a condensing agent) in a gas phase reactor, the product polymer is generated in solid particle form. These solid particles contain a sizable amount of condensing agent, monomer, and undesired species that need to be removed before the polymer can be sent to downstream units for processing and shipment. The removal of the hydrocarbons and impurities is done by counter-currently contacting the polymer particles with stripping nitrogen in a tower called the product purge bin, which can be challenging to model accurately.

This work focuses on the development of a practical model that can capture the major characteristics of the polyethylene product recovery process in the purge bin so they can be easily incorporated into Aspen Plus. The model can be applied to polyolefins (polyethylene, polypropylene) production to increase capacity, optimize performance, and guide plant operations to achieve sustainability targets.

454502 *Featured:* Organic Aqueous Flow Batteries for Massive Electrical Energy Storage

Abstract:

The ability to store large amounts of electrical energy is of increasing importance with the growing fraction of electricity generation from intermittent renewable sources such as wind and solar. Flow batteries show promise because the designer can independently scale the power (electrode area) and energy (arbitrarily large storage volume) components of the system by maintaining all electro-active species in fluids. Wide-scale utilization of flow batteries is limited by the abundance and cost of these materials. We have developed an approach to electricity storage using the aqueous redox chemistry of small, inexpensive organic molecules such as quinones [1,2]. The absence of active metal components in both redox chemistry and catalysis represents a significant shift away from the direction modern battery R&D has been taking. This new approach may enable massive electrical energy storage at greatly reduced cost.

[1] B. Huskinson, M.P. Marshak, C. Suh, S. Er, M.R. Gerhardt, C.J. Galvin, X. Chen, A. Aspuru-Guzik, R.G. Gordon and M.J. Aziz, â€œA metal-free organic-inorganic aqueous flow batteryâ€, *Nature* **505**, 195 (2014).

[2] K. Lin, Q. Chen, M.R. Gerhardt, L. Tong, S.B. Kim, L. Eisenach, A.W. Valle, D. Hardee, R.G. Gordon, M.J. Aziz and M.P. Marshak, "Alkaline Quinone Flow Battery", *Science* **349**, 1529 (2015).

458957 *Featured:* Hydrogen-Based Redox FLOW Batteries

Abstract:

Demand for large-scale energy storage is increasing, especially as renewable energy resources, such as solar and wind, become more prevalent. These variable energy generators require that large-scale energy storage is used to arbitrage and minimize fluctuations into the grid. As a promising candidate for energy storage and load leveling, redox flow batteries (RFBs) have been considered. However, due to the challenging issues such as low cell performance, durability, and high electrolyte cost, their wide-spread adoption has not occurred.

An electrochemical RFB system capable of meeting these demands is the hydrogen-bromine system. It has highly reversible and kinetically favored electrochemical reactions, soluble chemical species, high current-density operation, and a relatively inexpensive electrolyte. We have shown that the optimization and understanding of the primary losses can result in power densities of 1.5 W/cm^2 and limiting current densities over 4 A/cm^2 for discharge at ambient conditions. However, a key attribute of any RFB system is robust and durable performance, especially upon cycling. In this talk, we will report on the cyclic performance of the RFB, and especially the effect of operating conditions such as electrolyte concentration, cut-off potential, and current on the cycling performance.

In addition, other hydrogen chemistries will be examined including inexpensive (H_2-Fe-ion) and high-voltage (H_2-Ce) systems. The various performances and tradeoffs of these systems will be detailed. Finally, the various results will be put into context using experiments and simulations to elucidate the most critical issues in the field and outline areas of needed research.

460720 Achievable Alcohol Concentrations and Membrane Requirements for Artificial Photosynthetic System

Abstract:

Artificial photosynthesis of liquid fuels is a potential source for clean energy. Alcohols are particularly attractive products because of their high energy density and market value per amount of energy input. The major challenges in photo/electrochemical synthesis of alcohols from sunlight, water and CO_2 are low product selectivity, high membrane fuel-crossover losses, and high cost of product separation from the electrolyte. The maximum achievable concentrations of various alcohols in the catholyte are of the order of 1 mol % at partial current densities of 10 mA cm^{-2}. However such concentrations can only be achieved about 100 h of batch operation and result in in 100% fuel-crossover if a Nafion 117 membrane is used to separate the catholyte and anolyte compartments. Reducing the alcohol permeability of the ion-exchange membrane by an order of magnitude only increases the liquid fuel concentration to 10 mol%, but also raises the batch operation time needed to achieve this concentration to 1000 h. To overcome these limitations, we propose an artificial photosynthesis scheme for direct synthesis and separation to almost pure ethanol with minimum product crossover using saturated salt electrolytes. The ethanol produced in the saturated salt electrolytes can be readily phase separated into a microemulsion, which can be collected as pure products in a liquid-liquid extractor. A novel design of an integrated artificial photosynthetic system is proposed that continuously produces >90 wt% pure ethanol using a polycrystalline copper cathode at a current density of 0.85 mA cm^{-2}. The annual production rate of > 90 wt% ethanol using such a photosynthesis system operating at 10 mA cm^{-2} (12% solar-to-fuel (STF) efficiency) can be 15.27 million gallons per year per square kilometer, which corresponds to 7% of the industrial ethanol production capacity of California.

454317 Integrated Solar Energy Conversion and Storage Using Solar Flow Batteries

Abstract:

Photovoltaics are rapidly approaching cost parity with conventional fossil fuel sources of electricity, but they are hampered by the intermittent availability of sunlight. Electrochemical devices like batteries and fuel cells can facilitate large scale solar energy storage, but the capital intensity and long-term costs of current technologies remain prohibitive in most markets. We have made considerable progress toward realization of efficient solar conversion, storage, and delivery as electricity using an integrated photo-electrochemical device we call the Solar Flow Battery (SFB). We believe that a mature SFB technology will provide considerable cost advantages relative to discrete solar+battery systems due to the intensification of the overall energy conversion and storage process into a single device. In this presentation, recent results on SFB device modeling and realization of laboratory-scale prototypes will be discussed.

466161 Dynamic Energy Supply By a Pilot Scale Liquid Organic Hydrogen Carrier Unit

Abstract:

Liquid Organic Hydrogen Carriers (LOHCs) can store hydrogen in a chemically bound form at ambient conditions. If there is a need for energy, a dehydrogenation reactor is used to release hydrogen at temperatures at around 580 K. To function as a storage for electrical energy, the chemical reactor has to be supplemented by further equipment. In terms of efficiency heat recovery is crucial, because of the endothermal nature of the reaction. Furthermore, depending on the reconversion technology, hydrogen has to fulfil certain purity demands. Using a PEM fuel cell as a hydrogen reconversion system, its dynamics are limited by an insufficient supply with hydrogen in times of increased electric demand. There are two major process parameters that can be used to increase the hydrogen production rate in the dehydrogenation reactor.

Volume flow of the reactants: increasing the pump flow rate is a fast and dynamic possibility, but results in lower LOHC discharge levels and efficiency due to reduced residence time.

Temperature: increased temperature enhances the hydrogen release, as the reaction responds quite fast to temperature changes. However, dynamics in practice suffer from the thermal inertia of the system.

If hydrogen release and electrical power production do not match, the hydrogen pressure in the LOHC system is changing. Since there is a free gas volume between the fuel cell and the chemical reactor, the hydrogen gas phase can serve as an interim energy buffer. This free volume is constituted not only by the tubing, but also by the heat exchangers and hydrogen purification units. Thus, the LOHC system is able to cover dynamic electrical energy demand, which would exceed the dynamic capability of the dehydrogenation reactor.

To characterize the dynamics of the whole LOHC process, a pilot scale reaction system based on dibenzyltoluene was combined with a heat integration, a hydrogen purification system, a PEM fuel cell and different electricity consumers. In this contribution the dynamic performance of such a system is discussed.

474228 Fabrication of $CoTiO_3$/g-C_3N_4 Hybrid Photocatalysts with Enhanced H_2 Evolution

Abstract:

Hydrogen (H_2) has gained much attention as a sustainable and carbon-neutral fuel carrier. Since Fujishima and Honda reported H_2 evolution from water splitting by using TiO_2 electrode under light irradiation in 1972, photocatalytic water splitting has become one of the most fascinating and promising technologies. Among numerous photocatalysts, graphitic carbon nitride (g-C_3N_4) has been rapidly developed due to its non-toxic, proper energy band level for H_2 production (2.7 eV), ability of absorbing visible light, good physicochemical stability and facile fabrication via one-step polymerization. Unfortunately, the pure g-C_3N_4 seriously suffers from poor photocatalytic efficiency because of a high recombination rate of photo-generated charge carriers. To settle this problem, g-C_3N_4-based heterojunctions were prepared. However, the redox ability of photoexcited electrons and holes would be declined as the typical heterojunction forms.

This work reports a novel direct Z-scheme $CoTiO_3$/g-C_3N_4 (CT-U) photocatalytic system with different weight percentage of $CoTiO_3$ using a facile in-situ growth method for H_2 evolution from water splitting. The as-prepared CT-U composites composed of 1D $CoTiO_3$ microrod and 2D g-C_3N_4 nanosheet were characterized by various techniques including XRD, SEM, TEM, XPS, FTIR and UV-vis. Results demonstrate that the CT-U composite photocatalysts were successfully fabricated, with intimate interfacial contact and heterojunction interaction between g-C_3N_4 and $CoTiO_3$ which can significantly boost the photocatalytic activity compared with prinstine g-C_3N_4 and $CoTiO_3$. The most enhanced H_2 evolution rate of 858 Î¼molÂ·h^{-1}Â·g^{-1} and high quantum efficiency (38.4% at 365 nm, 3.23% at 420 Â± 20 nm) are achieved at an optimal 0.15% CT-U. Meanwhile, the 0.15% CT-U sample exhibits good

photocatalytic stability in recycling H_2 evolution. Accordingly, direct Z-scheme mechanism capable of leading efficient charge carrier separation and strong reduction ability for enhanced H_2 production was proposed, and further evidenced by PL, Photoelectrochemical analysis and ESR assay.

464431 Combining Thermochemical with Sensible Heat Storage for Increasing the Stored Energy Density and Providing Stable Heat Transfer Fluid Outflow Temperature

Abstract:

Concentrated solar power (CSP) plants with integrated thermal energy storage (TES) are considered to be a promising option for the cost-effective electricity generation and dispatchability because they allow the temporal decoupling of electricity generation and solar insolation. TES systems can be classified as sensible heat, latent heat, or thermochemical storage (TCS) systems [1]. In a TCS, the heat is looped via reversible chemical reactions: it is stored as the enthalpy of the endothermic reaction and recovered as sensible heat through the exothermic recombination of the reactants.

A competitive advantage of TCS compared to sensible or latent TES is the ability to control the charging and discharging temperatures through an adjustment of the equilibrium temperature of the reaction [2]. This requires a physical separation of the chemical reactants from the heat transfer fluid (HTF), e.g., by placing the participants of a gas-solid reaction in metal tubes. By connecting the tubes to an external gas compression unit, the partial pressure of the gas reactant can be continuously adjusted to maintain the reaction at the desired rate and temperature and thereby heat the HTF passing the tubes to the desired temperature.

The ability to control the temperature of the HTF passing a TCS section is expected to provide several advantages compared to current state-of-the-art sensible and latent heat storage systems:

- By heating the HTF to a constant outflow temperature during discharging, the power block can be operated at its design point and therefore at a high thermal-to-electricity efficiency.
- By cooling the HTF to a constant outflow temperature during charging that is equal to the temperature of the HTF leaving the power block, these two HTF streams can be merged and recirculated to the solar field without creating exergy losses due to mixing.
- By distributing multiple TCS sections along the TES axis, the axial temperature gradient can be increased, leading to increased energy storage densities.

To reduce the overall material costs of the TES, the TCS sections may be combined with sections comprised of packed low-cost sensible-heat storage (SHS) materials such as rocks or steel slag. If the HTF and the storage and encapsulation materials in the TCS and SHS sections are selected accordingly, the operating temperature can exceed the limit of ~600°C that is typically encountered in state-of-the-art two-tank TES systems.

Whereas the adjustment of the gas reactant pressure in a TCS offers some distinct advantages compared to other TES concepts, it introduces an additional energy penalty. Future work will focus on (1) reducing the losses due to adjusting the gas-reactant pressure, (2) identifying suitable TCS materials, and (3) optimizing the combined storage performance through a variation of the TCS materials, storage geometry and/or operating conditions.

References

[1] S. Kuravi, J. Trahan, D. Y. Goswami, M. M. Rahman, and E. K. Stefanakos, "Thermal energy storage technologies and systems for concentrating solar power plants," *Prog. Energy Combust. Sci.*, vol. 39, no. 4, pp. 285–319, Aug. 2013.

[2] Y. Q. Yu, P. Zhang, J. Y. Wu, and R. Z. Wang, "Energy upgrading by solid-gas reaction heat transformer: A critical review," *Renewable Sustainable Energy Rev.*, vol. 12, no. 5, pp. 1302–1324, 2008.

471609 Challenging Separations in the Power Generation Industry

Abstract:

The power generation industry has diverse challenging physical and chemical separations, including many in which membranes are important. For coal-fired steam generators, physical separations of ash (particulate matter) from the flue gas using electrostatic precipitation, wet scrubbing, or fabric filters pose challenges on various fuel types (interference from other constituents). Chemical separations include adsorption of contaminants such as mercury, reactive separations targeting sulur dioxide or sulfuric acid, and removal of carbon dioxide. Environmental contaminants (such as mercury, selenium, and nitrates) must also be removed from water used in the flue gas cleanup and ash handling.

However, other power generators include challenging separations. Energy storage technologies such as batteries and flow batteries rely on membranes to maintain charge separation and respond quickly when electrical demand is high. Advanced nuclear power generators need process contaminants removed to maintain high efficiency.

Poly-generation (such as hydrogen and power production) relies on separation technologies. Produced hydrogen must be separated from co-products using hydrogen separation membranes, and utilized in fuel cells that requires different membranes. Advanced hydrogen generation technologies such as electrolysis and photolysis require advanced membranes, and thermochemically produced hydrogen has challenging separations in each of the unit operations.

This talk will survey the key progress and remaining challenges in each area of power generation: coal-fired, storage, nuclear, and hydrogen.

473791 Self-Adaptive Water Treatment and Desalination of Source Water of Variable Feed Quality

Abstract:

Membrane based water treatment and desalination technologies have been gaining momentum in potable water production from seawater and brackish groundwater, municipal and industrial water reuse, as well as for management and reuse of agricultural drainage water. Considerable progress has been made since the early days of Reverse Osmosis (RO) water treatment/desalination regarding reduction in energy consumption, development of high performance fouling and chlorine resistant membranes, as well as improvements of in RO feed pretreatment. Given the increased level of water scarcity in various regions of the world and in particular in the Western US there has been growing efforts to expand the use of otherwise unutilized inland groundwater resources. A major challenge, however, is to implement a strategy that produces treated water of the right quality for the right use at the desired location. Temporal variability of source water quality, however, is a major challenge. Also, many of the various locations/communities, where water purification/desalination is needed, are remote and are not part of a centralized water infrastructure. In such locations, it is critical that water purification is accomplished at high recovery. Addressing the above challenges requires the development of a robust strategy for self-adaptive operation of membrane based filtration and desalination operations that will enable the deployment of distributed water systems. Such water systems have to readily adapt to changing water capacity demand and water quality fluctuations. Accordingly, an approach will be described whereby the concept of self-adaptive distributed water systems is developed, following a rapid process field evaluation approach that then serves to optimize process configuration. The above is enabled via direct real-time membrane monitoring and considerations of high recovery in a small plant footprint with salt harvesting via novel accelerated process of chemical demineralization. Examples of recent advances in the above areas will be discussed along with presentation of successful field demonstrations of seawater, brackish and industrial water treatment and reuse, in addition to purification of impaired groundwater.

473771 Desalination Performances of Large Hollow Fiber-Based DCMD Devices

Abstract:

Membrane distillation (MD) can be utilized to desalt highly saline waters that have considerable osmotic pressure. Current reverse osmosis desalination technology is not designed for saline waters having high osmotic pressure. Realization of this goal requires MD membranes to be highly fouling resistant. Fouling is a major obstacle that MD technique should potentially eliminate or control by selecting appropriate membrane material/surface, flow manipulation, additional force field and cleaning procedure. In direct contact membrane distillation (DCMD), Song et al. (2008) have shown that a rectangular cross-flow hollow fiber membrane configuration with a specific type of plasma polymerized superhydrophobic coating on the membrane outside surface and a novel method of operation ensured no precipitation based fouling from $CaCO_3$ and $CaSO_4$. We dwell on this novel aspect compared to conventional methods of fouling control. The simulation model of Song et al. (2008) assumed the membrane mass transfer coefficient, k_m, as an adjustable parameter for large hollow fiber modules with crossflow in a given range guided by estimated values and Zukauskas equation for predicting heat transfer coefficients in the two boundary layers. Here k_m was predicted from an enhanced model from Li and Sirkar (2016) without any adjustable parameters. To develop efficient designs, the model further simulated the performance of hollow fiber modules where the fiber ID was varied. Effects of fiber length on water vapor flux and water production rate were also modeled. This analysis will help design optimum hollow fiber membrane modules for desalination by DCMD for brines with low as well as high salinity.

472716 Development of a Multi-Stage Membrane Distillation System for Treatment of Hydraulic Fracturing Flow Back Waters

Abstract:

The reuse of wastewater for beneficial uses has become increasingly important in recent years. There is an urgent need to develop innovative and more effective technologies for treatment of wastewaters. Many of these wastewaters such as hydraulic fracturing flow back waters, contain very high dissolved solids (TDS). For example, treatment of hydraulic fracturing flowback waters can be very challenging as they often contain very high TDS, in excess of 200,000 ppm as well as dissolved surfactants and small organic compounds. Treatment of these highly impaired wastewaters is therefore very challenging. Pressure driven membrane processes such as reverse osmosis are impractical for treating very high salinity wastewaters due to the high osmotic back pressure that must be overcome. Membrane distillation has been proposed as a new unit operation for treatment of very high TDS wastewaters. One of the advantages of membrane distillation is the fact that low grade waste heat may be used.

Here we have developed a multi-staged membrane distillation system for maximizing water recovery from actual hydraulic fracturing flowback waters. The multi-stage system has been designed to different on-site applications for wastewaters with specific TDS levels. We show that ideal membrane properties vary as the TDS of the feed increases above 300,000. At lower TDS values membranes should be selected that maximize flux. However at TDS values above 300,000, membrane surface properties such as roughness becomes important. At theses higher TDS values leakage of the feed through the membrane pores and into the permeate side is a major concern. In addition appropriate pretreatment of the feed to remove dissolved organic compounds is essential prior to membrane distillation.

Here different membranes have been used in the different stages and performance of the system in terms of water recovery potential and specific energy consumption has been evaluated. Significant concentration of the feed has been conducted and over 60% water recovery has been achieved for actual hydraulic fracturing flowback waters from

Marcellus shale containing over 100,000 ppm TDS.

465789 Role of Membrane in Iemb Treatment of Groundwater Highly Contaminated with Oxyanions

Abstract:

Role of Membrane in IEMB treatment of groundwater highly contaminated with oxyanions

Perchlorate contamination of ground water is a worldwide concern. Several sites in Israel's coastal aquifer are contaminated with perchlorate, with hundreds of ppm found in one area accompanied with significant concentrations of nitrate and chlorate as well. This has prevented water production from wells in the area.

The ion-exchange membrane bioreactor (IEMB) [1] is a hybrid process for safe treatment of groundwater highly contaminated with oxyanions (perchlorate, nitrate, and chlorate). By Donnan dialysis anionic contaminants migrate across an anion exchange membrane (AEM) from a feed-water compartment to a bio-compartment. Once in the bio-compartment, the anions undergo microbial reduction to safer species such as chloride ions and nitrogen. The AEM acts as a barrier and keeps both compartments completely separate. Glycerol is used as an exogenous carbon and electron source for the biodegradation process [2]. This arrangement keeps the carbon source, reaction byproducts and bacteria confined in the bio-reactor thus preventing secondary contamination of the treated water. This contrasts with standard anaerobic bioreactors for treating such contaminations, for which the Ministry of Health requires additional barriers including MF, UV and GAC.

The present study examined the role of membrane type in performance of the IEMB in removing perchlorate and other anions (nitrate and chlorate) at levels of hundreds mg L^{-1} from ground water with a mixture of high levels of oxyanions. This was based on Donnan dialysis studies and was studied initially for synthetic and actual ground water fed to the water side while feeding 0.1 N of NaCl to the bio-compartment. To understand the results, conductivity and selectivity measurements were carried out on the different membranes. The anion exchange membranes examined included ACS (Tokuyama Soda), PCA-100 (PC-Cell GmBH), and AMI-7001 (Membrane International).

For ACS membrane and all experiment conditions studied here, perchlorate dominates the flux across the AEM and fluxes deviated from linear dependence on the driving force at less than 0.5 mM. Even though perchlorate concentration is considerably lower than nitrate and chlorate, its flux is greater than the sum of the other anion fluxes. At an effective perchlorate driving force (EDF) value above 0.7 [mM] perchlorate had a negative effect on the other anions resulting in a decrease in the flux for both nitrate and chlorate while its own flux only increased.

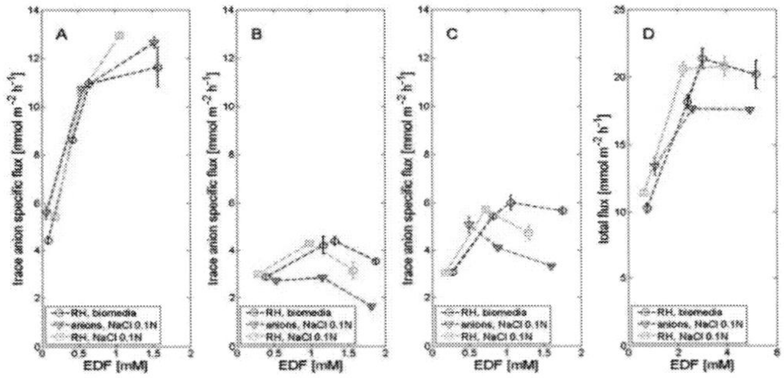

Figure 1 – Results of Donnan dialysis on ACS membrane. Oxyanions specific flux vs. the Effective Driving Force (EDF) of (A) perchlorate, (B) chlorate, (C) nitrate and (D) anions total flux vs. the total EDF, for IEMB Donnan dialysis experiments fed with (○) RHGW and bio media, (▼) synthetic anions solution and NaCl 0.1N and RHGW and NaCl 0.1N (■) into the water compartment and bio-compartment, respectively.

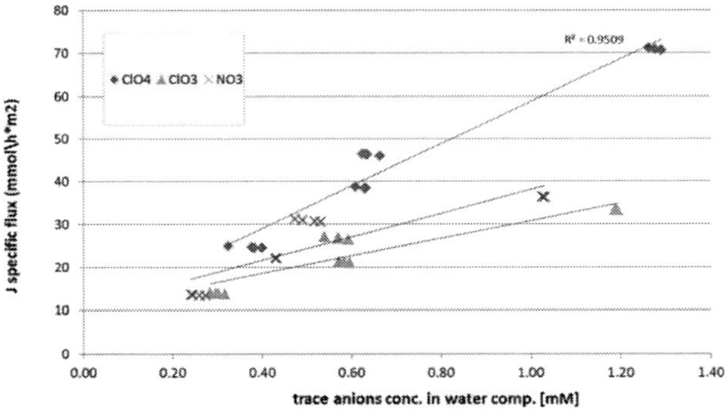

Figure 2: Specific flux in Donnan dialysis on synthetic mixture of oxyanions using PCA-100 anions exchange membrane. Bio-side contained 0.1 N NaCl.

On the other hand, experiments with PCA-100 membranes showed a linear dependence of flux on driving force to much higher levels and much less intereference of the perchlorate on the transport of nitrate and chlorate (Figure 2). Tests of membrane conductivity and regenerability supported the improved results with the PCA-100 membranes. Another type (AMI-7001) also showed promise based on conducicitivity but may be problematic in terms of polymer structure. The membrane selectivity for perchlorate and the other oxyanions over chloride was clearly lower for the PCA-100 membranes than ACS. A qualitative model will be proposed to explain these findings and relate it to the differences in ion exchange membrane structure.

If time allows, the implications of these findings for operating the IEMB as a plug flow reactor will be outlined.

[1] A.D. Fonseca, J.G. Crespo, J.S. Almeida, M.A. Reis, Drinking water denitrification using a novel ion-exchange membrane bioreactor, Environ. Sci. Technol. 34 (2000) 1557–1562.

[2] S. Fox, Y. Oren, Z. Ronen, J. Gilron, Ion exchange membrane bioreactor for treating groundwater contaminated with high perchlorate concentrations., J. Hazard. Mater. 264 (2014) 552–9.

470031 Economic Comparison of Pressure Driven Membrane Processes to Electrically Driven Processes for Use in Hydraulic Fracturing

Abstract:

Hydraulic fracturing has provided many new resources for the United States in oil and gas. These resources continue to be an economic driver going forward in many areas of the country. However, significant challenges remain with the environmental sustainability of this technology. One of the areas that is of particular concern is the re-use of the water that is used in fracking operations. The current method of dealing with this water, deep well injenction, takes water out of the eco-system and has been shown to cause earthquakes. This water has high levels of suspeneded and disolved organics as well as high levels of dissolved salts. Because of this the osmotic pressures of these streams are significantly more than seawater reverse osmosis in several instanced. Our group has looked at several possible alternatives of dealing with this water including reverse osmosis, nanofiltration, electrodialysis, and electrodeionization. All of these technologies have different advantages depending on the concentration of the water and ultimate re-use objective. This study has shown that electrodeionization is the most flexible technology, due to selective ion removal and high water recoveries, with the potential to work for both re-use and drinking production as well as acceptable economics at high and low ion concentrations. Experimental results as well as economic modeling will be presented in this talk.

454885 Sustainability Assessment and Life Cycle Inventories of Bio-Based Chemicals Using Performance Indicators

Abstract:

The implementation in practice of sustainable development by society, government, and industry has demonstrated that sustainable solutions possess the potential for obtaining economic benefits without damaging the environment and human health. However, it is necessary to evaluate and quantify the effects of these new practices in the environment. Therefore, a more comprehensive description of environmental, social, and economic effects when implementing sustainable practices, e.g., for the design of chemical manufacturing processes, is needed. This contribution describes the GREENSCOPE (Gauging Reaction Effectiveness for the Environmental Sustainability of Chemistries with a multi-Objective Process Evaluator) tool as a support for decision makers when developing, designing, and evaluating sustainable processes and products. In addition, a life cycle inventory for chemical synthesis routes and processes is obtained as one of the main features of GREENSCOPE. A life cycle inventory of a chemical manufacturing stage is an important step in the development of life cycle assessments, which determine environmental, economic, and social effects throughout a productâ€™s life cycle stages. The GREENSCOPE sustainability evaluation is performed by employing environment, economics, energy, and efficiency process indicators. A case study demonstration using GREENSCOPE during a bottom-up design of bio-based chemicals will describe which indicators have shown improvement and simultaneously find additional opportunities for enhancements. Finally, this contribution will present the GREENSCOPE tool as an approach to achieving life cycle assessments of bio-based chemical products during their manufacturing life cycle stages.

470414 Modeling for the Life Cycle Assessment of Dry Reforming Technology   Challenges and Lesson Learnt

Abstract:

The dry reforming reaction offers several advantages in fixation of carbon dioxide, a greenhouse gas. The process however, is dependent on choosing the right conditions and catalysts, which primarily govern the formation of syngas and byproducts. The quality of syngas produced also has a major role in the selection of subsequent products, where there may be the need for introducing hydrogen sources for hydrocarbons.

The choice of a system boundary and accounting for this process in terms of carbon emissions becomes very crucial, as the sources of carbon on one hand supplies mass, while the reactions can be endothermic, where external heat needs to be supplied and requires a source of energy. If fossil sources of energy are used, the net carbon balance tends to be negative. In this paper, the authors explore the possibilities of conventional and non-conventional energy sources to supplement the dry reforming process, and for targeted products in modeling the life cycle stage of CO_2 utilization via the dry reforming reaction.

466883 Performance Analysis of Chemical-Looping Fixed Bed Reactors Integrated in Combined Cycle Power Plants

Abstract:

Chemical-looping combustion (CLC) offers a promising method for combining electricity generation with in-situ CO_2 capture.[1,2] A metallic oxygen carrier is used to oxidize a hydrocarbon fuel in CLC, thus fuel and air are never mixed, and a pure stream of CO_2 can be separated after condensing water vapor. Thus, CLC has the potential to achieve higher thermal efficiency than most other CO_2 capture technologies. In this work, fixed bed CLC reactor configurations are studied, because their potential for seamless integration with combined cycle (CC) power plants and the capability to operate at high pressures.[3,4] Extensive analysis showed that the a key issue in the design of a fixed bed CLC reactor is the operating strategy that controls the temperature variations inside the reactor bed, especially during the exothermic oxidation step.[5–9] The dynamic nature of fixed bed reactors, as well as the many kinetically controlled reactions, necessitate the use of dynamic modeling to evaluate power plant performance, efficiency, stability and feasibility under transient operation.

This work presents a dynamic model for an integrated CLC-CC power plant and transient analyses of the integrated plant performance. A tool chain is presented for modeling the integrated CLC-based combined cycle power plant, coupling different equation-oriented platforms. Optimization of high-pressure fixed bed CLC reactor is performed to improve the efficiency of the downstream CC by maximizing the utilization of the gas turbine of the power plant. Optimal operating strategies are explored for the control of the temperature variations inside the reactor bed. The resulting time-varying gas stream exiting the CLC reactor is used by the gas turbine and steam cycle of a dynamic combined cycle power plant model. The power plant model is first validated against steady-state data from an existing natural gas-fueled, combined cycle power plant.[10] Then the combustion chamber of this plant is substituted by a CLC island operating optimally for improving the gas turbine performance. The transient variations of the integrated plant in terms of power, temperature, and pressure profiles are presented.

The CLC-CC power plant was estimated to operate at an average efficiency of 48.2%. This efficiency was accomplished by optimizing operational strategy for fixed bed reactors and the utilization of the exhaust streams of oxidation and heat removal. Dynamic simulation of the integrated CLC-CC plant showed that the combined cycle is only slightly affected by the batch nature of operation of the fixed bed CLC reactors. The CLC-CC plant was estimated to generate a relatively flat power output, especially from its gas turbine. Future work will focus on plant simulation under part-load operation and simulation of the integrated CLC-CC plant subject to time-varying loads. The developed simulation environment paves the way for conducting real-time optimization of power cycles with high-efficiency CO_2 capture technologies.

Acknowledgements

This material is based upon work supported by the National Science Foundation under Grant No. 1054718.

References

1. Kvamsdal, H. M., Jordal, K. & Bolland, O. A quantitative comparison of gas turbine cycles with CO2 capture. Energy 32, 10–24 (2007).

2. Chen, C., Han, L. & Bollas, G. M. Dynamic simulation of fixed bed chemical looping combustion reactors integrated in combined cycle power plants. Energy Technol. (accepted) (2016).

3. Han, L. & Bollas, G. M. Chemical-looping combustion in a reverse-flow fixed bed reactor. Energy 102, 669–81 (2016).

4. Zhou, Z., Han, L. & Bollas, G. M. Overview of Chemical-Looping Reduction in Fixed Bed and Fluidized Bed Reactors Focused on Oxygen Carrier Utilization and Reactor Efficiency. Aerosol Air Qual. Res. 14, 559–571 (2014).

5. Spallina, V. et al. Reactor design and operation strategies for a large-scale packed-bed CLC power plant with coal syngas. Int. J. Greenh. Gas Control 36, 34–50 (2015).

6. Noorman, S. & van Sint Annaland, M. Packed Bed Reactor Technology for Chemical-Looping Combustion. Ind. Eng. Chem. Res. 46, 4212–4220 (2007).

7. Fernández, J. R. & Abanades, J. C. Conceptual design of a Ni-based chemical looping combustion process using fixed-beds. Appl. Energy 135, 309–319 (2014).

8. Spallina, V. et al. Investigation of heat management for CLC of syngas in packed bed reactors. Chem. Eng. J. 225, 174–191 (2013).

9. Hamers, H. P., Gallucci, F., Cobden, P. D., Kimball, E. & Van Sint Annaland, M. A novel reactor configuration for packed bed chemical-looping combustion of syngas. Int. J. Greenh. Gas Control 16, 1–12 (2013).

10. Kehlhofer, Rolf; Rukes, Bert; Hannemann, Frank; Stirnimann, F. Combined-Cycle Gas and Steam Turbine Power Plants. 165–209 (2011).

472811 Chemical Looping Carbon Neutral and Carbon Negative Schemes for a Gas to Liquid Fuel Facility Thermodynamic, Techno-Economic and Experimental Analysis

Abstract:

Carbon Capture Utilization and Sequestration (CCUS) is also a grand challenge for modern chemical engineers. Many new technologies strive to implement either CO_2 capture or CO_2 utilization, but the two processes are rarely thought of as mutually inclusive. In order to have a meaningful impact on the mitigation of CO_2 emissions it is paramount that CO_2 capture and utilization or sequestration technologies are developed in conjunction with each other. On a commercial scale, a fossil fuel process involving unprecedented simultaneous capture and utilization of CO_2 can be transformational and disruptive in conventional CO_2utilization and carbon capture markets. The OSU carbon neutral and carbon negative schemes presented using the methane to syngas process for liquid fuels production as a model example.

The methane to syngas (MTS) process developed at The Ohio State University (OSU) can efficiently convert methane to liquid transportation fuels and high value chemicals by producing a syngas of high purity with a desired H_2:CO molar ratio in a single step from natural gas while avoiding the use of catalysts, steam, and molecular oxygen produced from an air separation unit. This unique feature of the MTS process can lead to a dramatic reduction of the capital cost required for, as an example, the gas to liquid (GTL) plant to as low as $65,100/(bbl/day) from $86,200/(bbl/day) required for conventional auto-thermal reforming (ATR) process. Such a substantial reduction in the capital cost, confirmed by a third party engineering firm WorleyParsons, is due to a >60% reduction in capital investment for the syngas production unit by the MTS process, as compared to ATR. It, thus, allows the MTS process to remain economically competitive even with low West Texas Intermediate crude oil prices (~$35/bbl for a natural gas cost of $3/MMBTU). This presentation initially presents will focus on presenting a detailed thermodynamic sensitivity coupled with a comprehensive techno-economic analysis (TEA) comparison to a conventional auto-thermal reforming technology for producing 50,000 bpd of liquid fuels from natural gas. This will be followed by the presentation of the design approach and philosophy used for developing a process performance model of the MTS technology and integrating it into a reference GTL facility, performing necessary heat and mass balance calculations, developing equipment configurations and lists, assessing capital and operating costs, and performing an initial and final economic evaluation to assess the costs and benefits of the MTS technology. Finally, the effect of application of the carbon negative and carbon neutral concept schemes to the developed TEA will be presented in terms of percentage improvement in carbon efficiency for liquid fuel production, comparison to the base MTS case TEA, variation of technology parameters which are important in terms of being de-risked and experimental validation of the carbon neutral and carbon negative schemes will be discussed

467227 Techno-Economic Analysis of Metal-Organic Framework Materials for Onboard, Light-Duty Vehicle Hydrogen and Natural Gas Storage

Abstract:

Metal-organic frameworks (MOFs) are a class of porous materials comprised of inorganic units bridged by coordinating organic linkers. Due to their high surface area and tunable binding energies, MOFs are promising materials for onboard, light-duty, automotive natural gas and hydrogen storage. A manufacturing cost analysis was conducted to understand the cost drivers for synthesis of Ni_2(dobdc), Mg_2(dobdc), $Zn_4O(bdc)_3$, and $Cu_3(btc)_2$ at automotive industry relevant production volumes (up to 2.5 Mkg/year) based on an engineering scale-up of demonstrated laboratory synthesis procedures. Conventional lab-scale synthesis of MOFs is carried out in excess organic solvents such as DMF, at elevated temperatures (~80°C-120°C), and are optimized for product purity rather than for yield. A number of extrapolations were assumed based on demonstrated high volume chemical manufacturing techniques and recent improvements in lab-scale synthesis. Results will be presented covering baseline costs using demonstrated solvothermal synthesis methods in organic solvents, aqueous synthesis as an alternative to the baseline synthesis in organic solvents, and liquid assisted grinding as a proposed alternative method to reduce solvent. Projected costs will be compared against U.S. Department of Energy targets for onboard hydrogen and natural gas storage, and sensitivity analyses will be presented to understand cost drivers. The results are expected to help identify a path towards low-cost, high volume MOF synthesis and to help guide the MOF research community towards avenues of investigation that lead to cost savings.

dobdc4 = 2,5-dioxido-1,4-benzenedicarboxylate

bdc2â€" = 1,4-benzenedicarboxylate

btc3â€" = 1,3,5-benzenetricarboxylate

474848 Effect of Confined Ion Cluster Scale on Catalytic CO_2 Conversion into Cyclic Carbonates

Abstract:

Effect of confined ion cluster scale on catalytic CO_2 conversion into cyclic carbonates

Suojiang Zhang, Qian Su and Weiguo Cheng

Key Laboratory of Green Process and Engineering, Institute of Process Engineering, Chinese Academy of Sciences, P.O. Box 353, Beijing, 100190, China

**Corresponding author:sjzhang@ipe.ac.cn*

In recent years, the atom-economy reaction of CO_2 with epoxides into cyclic carbonateshas attracted extensive attention in both academia and industry. Up to now, different forms of ionic liquids (ILs)have been developed for catalytic conversion of CO_2 into cyclic carbonates, such as bulk ILs, supported ILs and poly(ionic liquids).[1-4]Though a great improvement has been achieved in catalytic performance of these ILs system, which mostly consists of clusters with different ion numbers. What scale of ion cluster functions most efficiently on this transformation is still unknown and lacking of effective way to explore.

Confinement method provides possibility for this exploration.[5] By screening micro- and meso-porous materials with certain size of open channels as limited space, ion clusters of certain scale, ranging from one to dozens of angstroms, could be confined and isolated via in-situ synthesis in each space. In this work, molecular sieves were adopted as limited spaces for investigation about the effect of ion cluster scale on CO_2 conversion into cyclic carbonates. Through

experiment test and molecule simulation, the space size of molecular sieves was rigorously screened to fix ion clusters firmly in and allow reactant and target product molecules free in-and-out. By using different size of spaces, different scale of clusters with varying ion numbers could be confined and the corresponding catalytic activity along with scale change, especially from one to dozens of angstroms, would be evaluated. This study not only optimizes the ion cluster scale which is needed for efficient synthesis of cyclic carbonates, but also provides an efficient way to investigate angstrom scale reaction mechanism, which is of great significance in both academic and industrial areas.

Keywords: ionic liquids, CO_2, cyclic carbonates, ion cluster, scale

References

1.Zhang SJ, Sun J, Zhang XC, Xin JY, Miao QQ, Wang JJ. Ionic liquid-based green processes for energy production. *Chemical Society Reviews*, 2014, 43: 7838-7869

2. Cheng WG, Chen X, Sun J, Wang JQ, Zhang SJ.SBA-15 supported triazolium-based ionic liquids as highly efficient and recyclable catalysts for fixation of CO_2 with epoxides. *Catalysis Today*, 2013,200: 117-124

3. Shi TY, Wang JQ, Sun, J, Wang MH, Cheng WG, Zhang SJ. Efficient fixation of CO_2 into cyclic carbonates catalyzed by hydroxyl-functionalized poly(ionic liquids). *RSC Advances*. 2013,3: 3726-3732

4.Xu BH, Wang JQ, Sun J, Huang Y, Zhang JP, Zhang SJ. Fixation of CO_2 into cyclic carbonates catalyzed by ionic liquids: a multi-scale approach. *Green Chemistry*, 2015, 17: 108-122.

5. Ban YJ, Li ZJ, Li YS, Peng Y, Jin H, Jiao WM, Guo A, Wang P, Yang QY, Zhong CL, Yang WS. Confinement of ionic liquids in nanocages: tailoring the molecular sieving properties of ZIF-8 for membrane-based CO_2 capture. *Angewandte Chemie*. 2015, 127: 15703-15707

475358 Isopentane and Ethylene Alkylation Using Ionic Liquid Catalyst

Abstract:

Modern refineries employ many upgrading units such as fluid catalytic cracking (FCC), hydrocracking (HCR) and PenHex paraffin isomerization. As results, these refineries produce a significant amount of isopentane. Historically, isopentane was a desirable blending component for gasoline having higher octane (92 RON), although it exhibited high volatility (20.4 Reid vapor pressure, RVP). As environmental laws began to place more stringent restrictions on gasoline volatility, the use of isopentane in gasoline was limited because of its high volatility. As a consequence, the problem of finding uses for by-product isopentane became serious, especially during the summer season. Moreover, as more gasoline composition contain ethanol as their oxygenate component, more isopentane must be kept out of the gasoline pool in order to meet the gasoline volatility specifications.

We explored upgrading of isopentane by alkylation with either ethylene or propylene to produce C7 or C8 alkylate using ionic liquid catalysts. Higher reactivity of the ionic liquid catalyst achieved nearly complete conversion of ethylene. This was surprising initially since conventional alkylation catalysts (HF or H2SO4 acid catalysts) cannot alkylate ethylene. We have studied impact of process variables for alkylate product quality and identified process conditions for scale-up. Continuous microunit runs were conducted to examine the impact of extended operation on catalyst aging.

We also examined potential low cost source of ethylene from a refinery. A dilute stream of ethylene is available from the FCC offgas. Currently the FCC offgas is sent to furnace for steam generation. By consuming the ethylene in the FCC offgas, refinery fuel gas yield would be lowered and high-value alkylate product can be created. This concept was particularly attractive since we are combining two low value refinery streams to produce clean, low RVP C7 gasoline. Integrated process concept for this technology was generated. C2 stream separation from the FCC offgas is a costly step in the process, and we are looking into ways to lower the overall process cost.

This talk will present the catalysis of isopentane alkylation and development of the technology.

474845 Dissolution, Separation and Conversion of Biomass with Ionic Liquids

Abstract:

Dissolution, separation and conversion of biomass with ionic liquids

Suojiang Zhang

Institute of Process Engineering, Chinese Academy of Sciences, Beijing 10080, China

E-mail: sjzhang@ipe.ac.cn

With the growing pressure on resource depletion and environment degradation, the development of green processes for productions of renewable energy and chemicals are highly required. Recently, we mainly focus on insights into green processes on dissolution, separation and conversion of biomass based on ionic liquids (ILs) and several the latest green processes are developed by using ILs. A two-step ionic liquids method for cornstalk pretreatment was developed and its detailed micro mechanisms were studied. After the pretreatment, the crystalline structure of cellulose was altered, amorphous cellulose was obtained, most of lignin was removed and cellulose content was increased substantially. Moreover, conversion of biomass-derived Î³-valerolactone into high quality gasoline was achieved by decarboxylation of valerolactone to produce butenes and alkylation of the produced butenes with butane using functional ionic liquid as an efficient catalyst. The obtained gasoline was rich in trimethylpentane with a high research octane number of 95.4.

However, the challenges are still great for developing industrial scale processes of dissolution and conversion with ionic liquids. More systematic work should be carried out deeply and broadly in the future.

Acknowledgement:

This work was supported by International S&T Cooperation Program of China (2014DFA61670), External Cooperation Program of BIC, Chinese Academy of Sciences (GJHZ201306), National Natural Science Foundation of China (No. 21276260, 21576269, 21476245, 21406230).

Reference:

[1] Zhang, S.; Sun, J.; Zhang, X.; Xin, J.; Miao, Q.; Wang, J. *Chem. Soc. Rev.* **2014**, *43*, 7838.

[2] Xin, J.; Yan, D.; Ayodele, O.; Zhang, Z.; Lu, X.; Zhang, S. *Green Chemistry* **2015**, *17*, 1065

[3] Xu, J.; Yao, X.; Xin, J.; Lu, X.; Zhang, S. *Journal of Chemical Technology & Biotechnology* **2014**, DOI 10.1002/jctb.4517

[4] Chen, L.; Xin, J.; Ni, L.; Dong, H.; Yan, D.; Lu, X.; Zhang, S. *Green Chem.* **2016**

455044 Ionic Liquid-Based Aqueous Biphasic Systems for High-Performance Extraction

Abstract:

Aqueous biphasic systems (ABSs) offer a benign alternative process for conventional extraction systems with volatile organic solvents to separate biomasses, such as amino acids, antibiotics, alkaloids, and proteins et al. And the employment of ionic liquids (ILs) in ABSs vastly promotes the extraction efficiency of various solutes because when compared with traditional polymers, ILs are highly designable and have tunable polarity in dealing with specific tasks. However, it is still challenging to construct an IL-based ABS with fine phase splitting ability, excellent extraction efficiency and low IL loss. Herein we report a series of newly designed IL-based ABSs that contains long-chain carboxylate anions and a cholinium cation that are all derived from biomass. This strategy introduced long alkyl chains into the anions, which not only significantly increased the hydrogen bond (HB) acceptor ability of the carboxylate anions through remarkable electron-donating effect but also ensured good hydrophobicity for achieving better phase splitting. The developed IL-based ABSs demonstrated a relatively broad biphasic area and extraordinary extraction efficiency for amino acids and bioactive compounds with partition coefficients for tryptophan, phenylalanine and caffeine of 58.54, 120, and 120, respectively, which were remarkably higher than those obtained in ABSs with conventional ILs. In addition, liquid crystal structures were observed when the carbon number of the carboxylate anion of the ILs exceeded eight; thus, IL-based ABSs with liquid crystal structures were reported for the first time. Furthermore, we synthesized a family of polymeric ionic liquids (PILs) to explore IL-based ABSs further. Phase diagrams determined by cloud point method demonstrated that the PILs have far more powerful ability to construct ABSs than conventional ILs. And it was demonstrated that in the designed PIL-based ABSs, unprecedentedly high partition coefficients were obtained (K_{Trp} = 1127, K_{Phe} = 188). Prospectively, PIL-based ABSs successfully combine the unique properties of PILs and ABSs, and offer a new platform for the development of green and high-performance extraction methods.

461755 Corn Stover Deconstruction Using Ionic Liquid â€“ Glycerol Mixtures

Abstract:

Intertwined cellulose, hemicelluloses, and lignin biomacromolecules make up the recalcitrant structure of lignocellulosic biomass. The configuration of this kind of biomass makes deconstruction to its components challenging. Yet the prospect of using biomass, particularly agricultural residues, for renewable energy or for sustainable chemicals is inviting. Corn stover, in particular, is available in the US Midwest in quantities estimated to be greater than 30 million metric tons/year.

Certain ionic liquids (ILs), salts that are molten below 100 °C, can preferentially dissolve lignocellulosic biomass fractions. The ILs studied in this work were 1-ethyl-3-methylimidazolium acetate and 1-ethyl-3-methylimidazolium formate. Though they are effective, do not produce hazardous emissions, and are recyclable, these ILs are costly. To reduce solvent expense, we examined diluting them with non-hazardous compounds. We found that glycerol, a low-value by-product of the growing biodiesel industry, is fully miscible with the selected ILs, and is a potentially useful co-solvent. Solvent mixtures, produced from 50% or 75% glycerol with the remainder being one of the selected ILs, were used to pretreat corn stover. Dilution of the ILs with up to 75% glycerol was found to enhance the glucose and xylose yields after enzymatic hydrolysis, compared to raw corn stover or corn stover treated with pure IL. FTIR spectra of the raw and pretreated corn stover suggests that lignin had been dissolved into the IL- glycerol mixture and that some of the crystalline cellulose had been converted to amorphous cellulose, which is more easily hydrolyzed.

468784 Side-Chain Cleavage of Phytosterols By *Mycobacterium* sp. in a Biphasic Ionic Liquid-Aqueous System

Abstract:

Androst-4-ene-3,17-dione (AD) is an important steroid intermediate which can be used for synthesis of various hormone drugs. Phytosterols, that are abundant in oil residues and papermaking waste liquor, are preferred raw materials for production of AD. However, low substrate solubility and product inhibition are two major problems existed in phytosterols biotransformation. Recently, biphasic systems are considered to be efficient strategies to solve above problems, in which the hydrophobic phase act as a â€œreservoirâ€ for substrate dissolving and in-situ product removal (ISPR).

Ionic liquids (ILs), with advantages of tunable physical and chemical properties, negligible vapor pressure, and non-flammability, are promising alternatives to be applied in biphasic systems. In this work, nine kinds of water-immiscible ILs were used in phytosterols side-chain cleavage by growing cells of *Mycobacterium* sp. MB 3683. The solubilities of phytosterols and AD measured in these ILs were greatly higher than those in water, and the logarithm of distribution coefficients of AD between ILs and water were all above 2.0, suggesting their potential for ISPR. Specifically, [PrMIM][PF_6] was determined as a suitable candidate to establish a biphasic IL-aqueous system for phytosterols biotransformation to AD, and the effects of inoculation, IL addition time, and phase ratio were studied in detail. The results indicated that IL preferred to be added to fermentation medium when the cells were in stationary phase due to the toxicity of IL on cell membrane, and a small quantity of IL was appropriate because of its superior dissolving property. Under optimal conditions, i.e. inoculum size of 12% with 48 h age, IL addition at 84 h, and phase ratio of 20:1 (v/v, aqueous/IL), AD production achieved 2.23 g L^{-1} after 5 days biotransformation with substrate concentration of 5 g L^{-1} in a scaled-up biphasic system of 200 mL. Moreover, IL recovery was attained to be above 90%, and [PrMIM][PF_6] can be sequentially used for five batches with stable conversion in AD production.

Owing to increasing environmental constraints from organic solvents, biphasic IL-aqueous systems which presented efficient conversion over monophasic aqueous system at high feeding concentration, are potential in microbial side-chain cleavage of phytosterols and up-scale AD production.

471288 Lipid Aggregation Thermodynamics and Kinetics in Ionic Liquid/Co-Solvent-Based Extractions

Abstract:

The extensive use of ionic liquids (ILs) is a significant achievement in extracting lipids and proteins from biomass for bio-fuel production from algae and other bio-based feedstocks. The particular effectiveness of ionic liquids combined with polar-covalent molecules (PCM), such as methanol, as an extraction solvent mixture in extracting bio-oils from micro-algae has been demonstrated successfully, the main highlight being auto-partitioning of extracted lipids into a separate phase.[1] While the efficacy of these IL/PCM mixtures has been demonstrated experimentally, our previous efforts focused on understanding the thermodynamic and structural properties of the basic co-solvent mixtures.

Experimental results showed that intermediate compositions of IL/PCM mixtures are desired for maximum lipid extraction yields.[2] Molecular dynamics (MD) simulations are conducted to predict the lipid aggregation behavior of triolein, a model lipid compound, at various concentrations of IL/PCM mixtures. MD generated radial distribution functions (***rdfs***) and lipid clustering analysis provide an estimate of the thermodynamics and kinetics of lipid aggregation and are compared to light scattering data. Clustering analysis and simulation snapshots indicates that the lipids are aggregating in these IL/co-solvent mixtures, a simulation finding that is supported by light scattering experiments and correlates well with auto-partitioning. A correlation between the extent of triolein aggregation and the experimental extraction yields as a function of co-solvent composition is also explored.

These efforts provide an improved understanding between the molecular-level behavior and macroscopic thermodynamic properties of lipid aggregation and auto-partitioning in IL/PCM mixtures, to characterize these new tailored solvent mixtures for bioprocessing separation applications.

References

[1] G. Young, F. Nippgen, S. Titterbrandt, and M. J. Cooney, "Lipid extraction from biomass using co-solvent mixtures of ionic liquids and polar covalent molecules", Separation and Purification Technology, **2010**, 72, 118.

[2] G. Severa, G. Kumar, M. Troung, G. Young and M. J. Cooney, (2013). "Simultaneous extraction and separation of phorbol esters and bio-oil from Jatropha biomass using ionic liquid–methanol co-solvents", Separation and Purification Technology, **2013**, 116, 265.

469447 Phase Behavior and Separations Using Ionic Liquids

Abstract:

The history of ionic liquids (ILs) effectively started in 1914, when the physical properties of ethylammonium nitrate ($[CH_3CH_2NH_3^+][NO_3^-]$, m.p. 13-14 °C) were first reported. ILs are generally defined as salts composed of discrete cations and anions with melting points below 100 °C, and many are liquid at ambient temperature. IL research has grown rapidly over the past decade due to the realization that these materials have many unique properties such as negligible vapor pressure and outstanding solvation potential. Ionic liquids have been further emphasized by the fact that their physical and chemical properties can be finely tuned by varying both the cation and anion. Our research has focused on accurately measuring vapor-liquid equilibria (VLE) and vapor-liquid-liquid equilibria (VLLE) and using thermodynamic models to understand the phase behavior of binary gas mixtures in ILs. This presentation will focus on the importance of characterizing the global phase behavior of gases in ionic liquids and how this can provide insight into new applications. Solubility measurements of several gases in ILs will be discussed and important experimental details regarding VLE measurements using a gravimetric microbalance and VLLE measurements using a mass-volume technique will be highlighted. VLE data have been successfully correlated with a modified Redlich-Kwong equation of state (EOS), and in certain cases (*e.g.* hydrofluorocarbons) the EOS predicts partial immiscibilities (LLE) with lower critical solution temperatures (LCSTs) in the fluorocarbon-rich side solutions. We have also found that gases such as CO_2 can exhibit different solubility behaviors in ILs (*i.e.* physical and chemical absorption) and that these behaviors can be analyzed with the EOS using a simple association model and excess thermodynamic functions. Knowledge of gas and IL phase behavior has led to several practical applications including separation of azeotropic gas mixtures, removal of CO_2 from flue gas and new fluid pairs for absorption cooling. These along with other interesting examples from the literature will be discussed.

473813 Z-Bond: New Insight into Ionic Liquid from Structure to Application

Abstract:

Z-Bond: New Insight into Ionic Liquid from Structure to Application

Kun Dong, Suojiang Zhang[1]

Beijing Key Laboratory of Ionic Liquids Clean Process, State Key Laboratory of Multiphase Complex System, Institute of Process Engineering, Chinese Academy of Sciences, Beijing 100190, China.

Keywords: Ionic Liquids, Z-bond, Properties, Applications

Ionic liquids have attracted intensive attentions in academia and industries due to their unique properties and potential applications. Nowadays, much interest has focused on the quantitative relationship between structures and properties. A most critical question is â€œwhat is the force to determine the properties of ionic liquids.â€ Unlike the NaCl in which the Coulomb force is solely dominant, in ionic liquids the H-bonds as the structural directing contributor, are responsible for ionic pairing, stacking and self-assembling, and then influence their properties. [1-2]

By DFT calculations combining with the experimental spectroscopy, we found that coupling with the strong electrostatic force, H-bonds in ILs are very different from the common solvents, such as water. We defined the interaction as Z-bond. [3] More interestingly, the Z-bonds can be extended into a 3D network that is the essential interactive fashion and structural units and determine the many properties of ILs. [4]

5

Moreover, involving in some important applications of ILs, the Z-bonds also play the critical roles. _ The typical example is the dissolution of cellulose in ILs that involves in the two steps H-bonded interactions essentially. The first is to break the H-bonded network inside IL, then the active anion can interact with the hydroxyl protons of cellulose by forming Z-bonds. Another, such as CO_2/SO_2 gases absorption, the natural bioactive homologous separation, as electrolyte in lithium batteries, even as catalytic in reactions, involve in the roles of H-bonds. Therefore, the deep insights into structures, in particular the Z-bonds, can provide us with a rational design for the new ILs to fulfill the demands in some complicated chemical processes.

References:

(1) Dong, K.; Zhang, S.; Wang, J. *Chem. Commun.* **2015**.

(2) Dong, K.; Zhang, S. *Chem. Eur. J.* **2012**, *18*, 2748.

(3) Dong, K.; Wang, Q.; Zhang, S. *Sci. China Chem.* **2015**, *58*, 495.

(4) Fumino, K.; Peppel, T.; Geppert-RybczyÅ„ska, M.; Zaitsau, D. H.; Lehmann, J. K.; Verevkin, S. P.; KÃ¶ckerling, M.; Ludwig, R. *Phys. Chem. Chem. Phys.* **2011**, *13*, 14064.

(5) Zhang, S.; Sun, J.; Zhang, X.; Xin, J.; Miao, Q.; Wang, J. *Chem. Soc. Rev.* **2014**, *43*, 7838.

---

[1] The corresponding author: sjzhang@home.ipe.ac.cn

464895 Thermal and Thermodynamic Properties of Ionic Liquids and Molten Salts with High Thermal Stability

Abstract:

Ionic liquids have attracted much attention in the past few decades because of their potential as novel process fluids. Their vanishingly low vapor pressures coupled with the ability to tune their properties through the full pallet of organic chemistry have made them promising alternative replacement solvents for volatile organic chemicals used as reaction and separation solvents. Additionally, their oft reported high thermal stability has made them potential candidates for high temperature reaction and separations solvents as well as heat transfer fluids. However, as many of the ionic liquids begin decomposing between 200 and 300°C, their use as high-temperature fluids is somewhat limited.

To this end, we have synthesized and studied a series of ionic liquids and molten salts that have shown thermal stability up to and above 300°C. We indicate molten salts in addition to ionic liquids as several of the salts have melting point above 100°C, the commonly used upper limit for the term ionic liquid. By judicious selection of cation and anion pairs, the thermal stability of these species can be significantly enhanced.

The species examined include a variety of cations, including peraryl phosphonium and sulfonium cations paired with several anions. In this work we present the melting points of these compounds, their solid and liquid heat capacities (measured via differential scanning calorimetry), as well as their thermal decomposition temperatures (measured via TGA). Additionally, we examine the relative contributions of enthalpy and entropy of fusion on the melting points of homologous series of the compounds. Furthermore, long-term, high-temperature stability studies of several of the species will be presented.

These initial studies will provide a basis by which to tailor additional ionic species and broaden the types of species available for use and study, and guide future work related to energy storage and high-temperature reactions in these non-volatile species.

465186 Multiscale Coarse-Graining of Ionic Liquid Electrolytes to Deliver Accurate Dynamics and Transport Properties at the Mesoscale

Abstract:

The air transportation system is a major part of the United States and global economies. Electric aircraft that is characterized with high energy efficiency, low emissions, and reduced noise is proposed to make the nation's air transportation system more efficient, safe, and sustainable. The application of new electrode materials together with alternative electrolytes based on ionic liquids (IL) have the potential to enable safe and high energy batteries. Ionic liquids are very attractive candidates for battery electrolytes because they have low volatility, moderate reactivity, low flammability, and a wider liquid range than most organic solvents. Computer modeling and simulation of the ionic liquid systems with the molecular dynamics (MD) approach yields an accurate prediction of the systems' structure, dynamics, and thermodynamic properties. However, application of the MD to the IL systems at mesoscale is limited and in most cases is prohibited due to extremely high computational cost. As an alternative, the coarse-grained molecular dynamics (CGMD) approaches may be used. In CGMD the number of degrees of freedom in the systems is significantly reduced by combining multiple atoms into a single coarse-grain (CG) particle. These approaches are successfully used for an accurate prediction of structure and thermodynamic properties. However, CGMD methods are not widely applied to the IL systems due to lack of accurate dynamic properties prediction without additional treatments. In this work a new Probability Distribution Function Coarse Grain (PDF-CG) method is applied to the system of ionic liquids to recover its dynamics properties. It is shown that application of the PDF-CG method accurately captures dynamics of the IL system in addition to the accurate prediction of the structure and thermodynamic properties. PDF-CG method may be successfully applied to the systems where accurate representation of dynamics is essential to advance computational capability up to the mesoscale.

NTNU

Norwegian University of Science and Technology

Prediction of CO_2 physical solubility in ionic liquids using artificial neural networks and COSMO σ-moments

Bouza, Pablo[1]; Díaz, Ismael[2]; Rodriguez, Manuel[2]; Palomar, José[3]; Haug-Warberg, Tore[1]

AIChE Meeting, November 13-18 2016
San Francisco, USA

[1] Department of Chemical Engineering, NTNU, Norway
[2] DIQUIMA, UPM, Spain
[3] Sección de Ingeniería Química, Universidad Autónoma de Madrid, Spain

1

NTNU

Background and Motivation

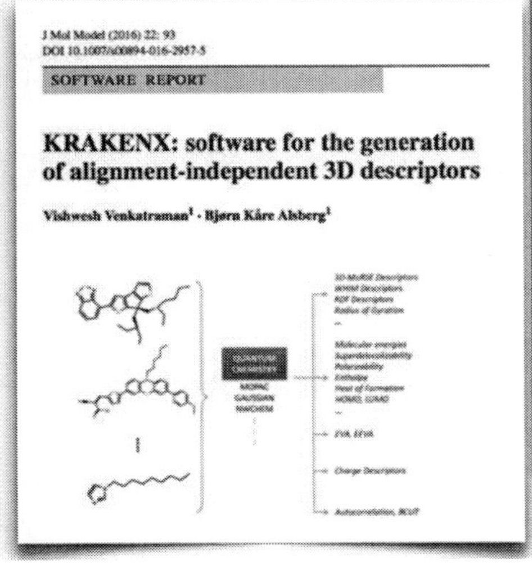
J Mol Model (2016) 22: 93
DOI 10.1007/s00894-016-2957-5

SOFTWARE REPORT

KRAKENX: software for the generation of alignment-independent 3D descriptors

Vishwesh Venkatraman[1] · Bjørn Kåre Alsberg[1]

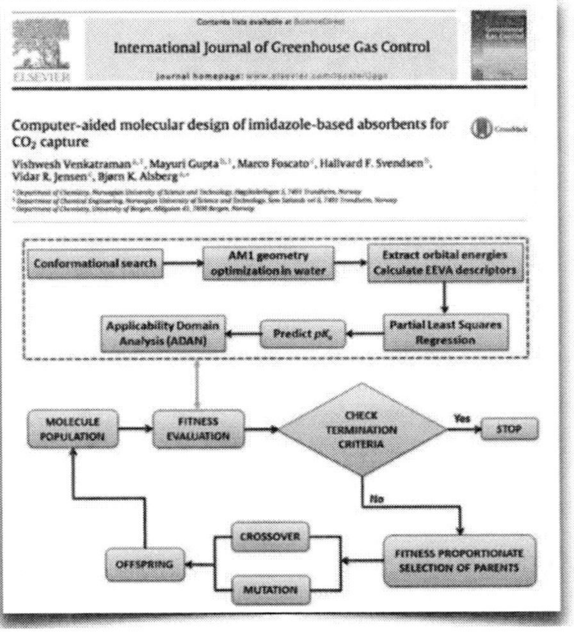
International Journal of Greenhouse Gas Control

Computer-aided molecular design of imidazole-based absorbents for CO_2 capture

Vishwesh Venkatraman, Mayuri Gupta, Marco Foscato, Hallvard F. Svendsen, Vidar R. Jensen, Bjørn K. Alsberg

2

3

Aims and Objectives

- Target property: **CO_2 solubility** in Ionic Liquids as function of T and P
- **Computational budget**
  - **Descriptor Calculation** max. 1 geometry optimization + single point calculation at DFT level per ion; **MD approaches ruled out**; Single conformer
  - **Evaluation** must $\in O(1)$, to enable high-throughput screening
- Methodology: nonlinear **QSPR**
  Quantitative Structure-Property Relationship
- Feature selection: **COSMO based descriptors**
- **Dataset**: most heterogeneous, well curated and current we can find
- **Benchmark** it with state-of-the-art methods
- **Establish its applicability domain** with a modern methodology.

4

Prior work

$$K_H = \gamma_i^{\infty} \cdot P_{vap}^{*}$$

- Both pure COSMO-RS and ANN on the σ-profile
- Related K_H to the excess enthalpy of dissolution (exothermicity of mixture)

\+ Successfully predicted Henry's constants for several gases in ILs

\+ Reproduces the effect of polyhalogenation in K_H and general anion trends.

\- Small train+test dataset (24 IL)

\- No pressure and temperature effects

\- Lacked explicit applicability domain definition

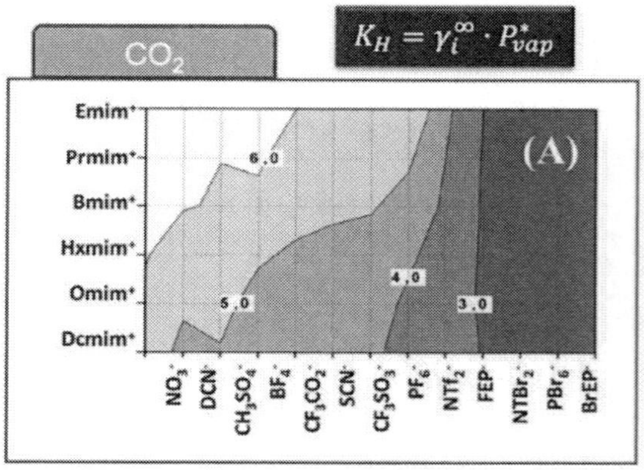

J. Palomar, M. Gonzalez-Miquel, A. Polo, F. Rodriguez. Understanding the physical absorption of CO2 in ionic liquids using COSMO-RS method. Industrial & Engineering Chemistry Research 2011, 50(6), 3452-346

NTNU

Workflow

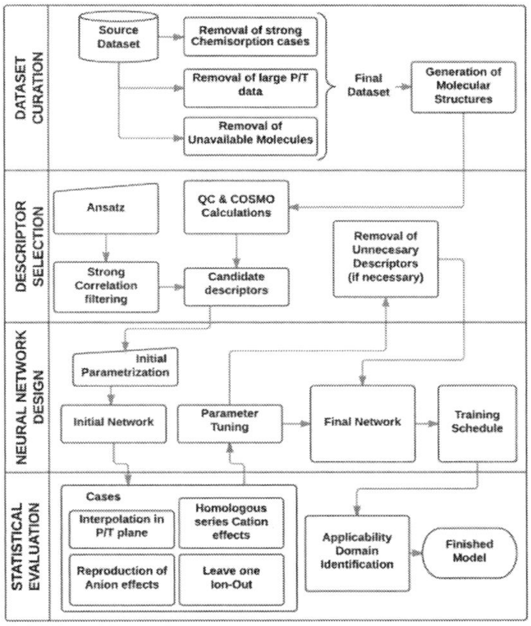

P. Bouza; I. Díaz; J. Palomar; V. Venkatraman; B. K. Alsberg.;T. Haug-Warberg. Prediction of CO_2 physical solubility in ionic liquids using artificial neural networks and COSMO σ-moments.

5

NTNU

Data Sources

- 9000 data points
- 136 kinds of ILs
- Includes COSMO-SAC and UNIFAC benchmarks

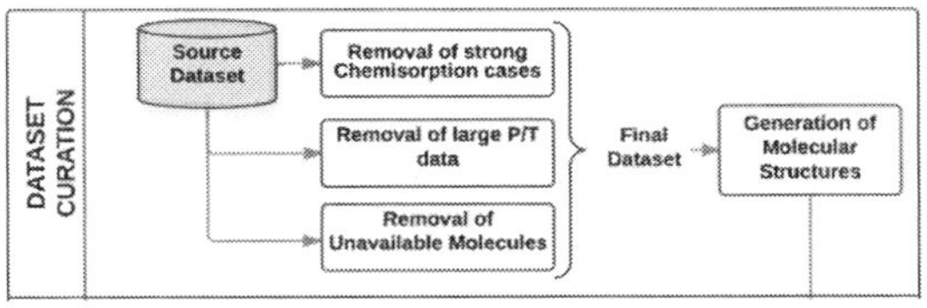

Frequently updated

Pure ionic liquid systems:
1138

Data points (pure ionic liquids):
85379

Properties:
Thermodynamic, transport; pure component, binary, tertiary mixtures;VLE,LLE, etc..

Drawback:
Web service only

6

Data Curation

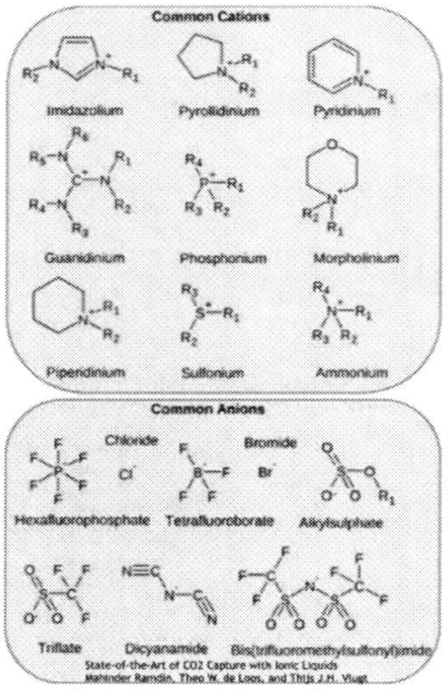

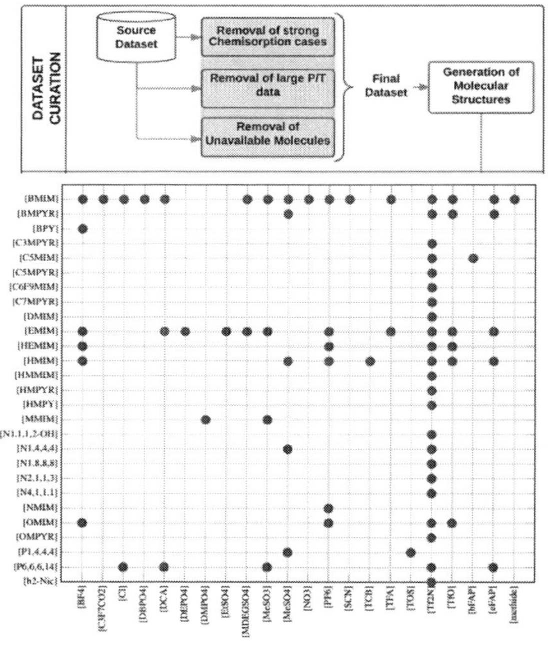

7

Data Curation

- Chemisorption: Acetates, sulfonates, etc

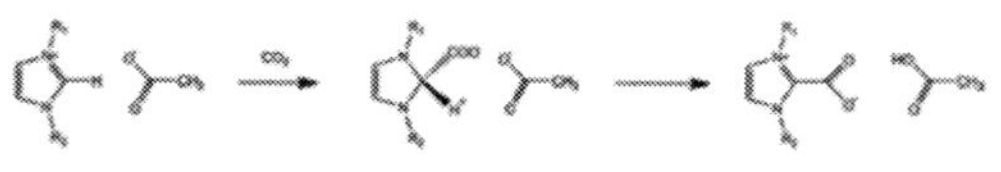

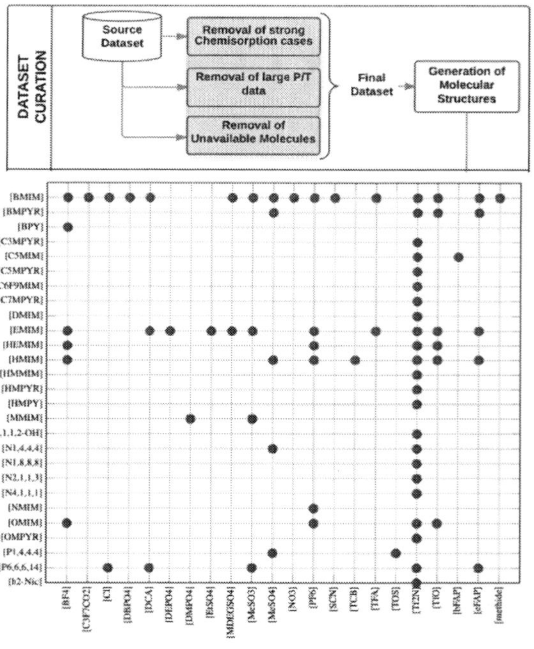

7

Data Curation

- Chemisorption: Acetates, sulfonates, etc
- Some of the datapoints are outside of practical P,T ranges

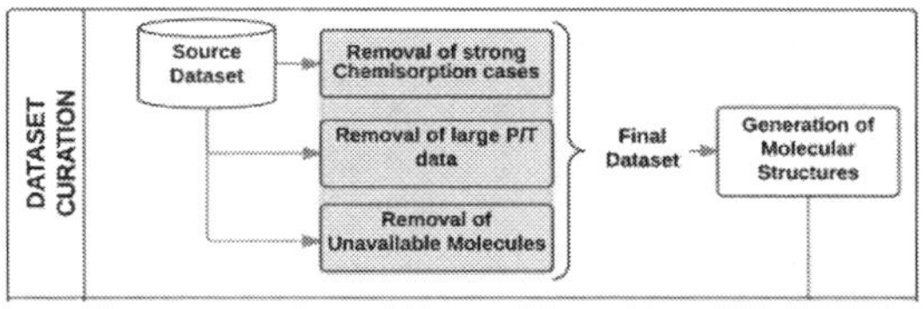

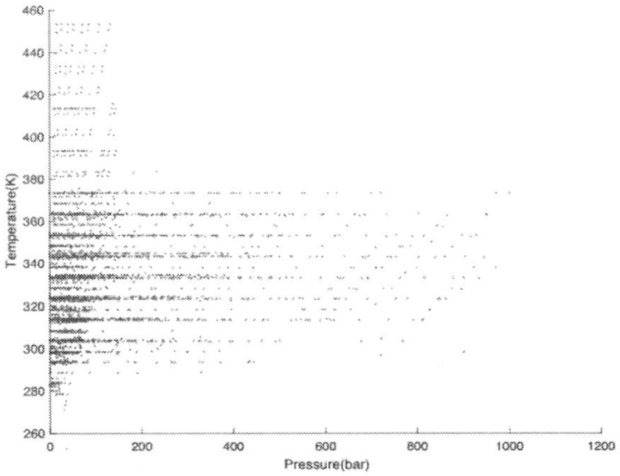

7

Data Curation

- Chemisorption: Acetates, sulfonates, etc
- Some of the datapoints are outside of practical P,T ranges
- Some ILs are problematic to model (ex. polymeric ILs)

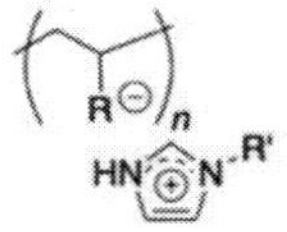

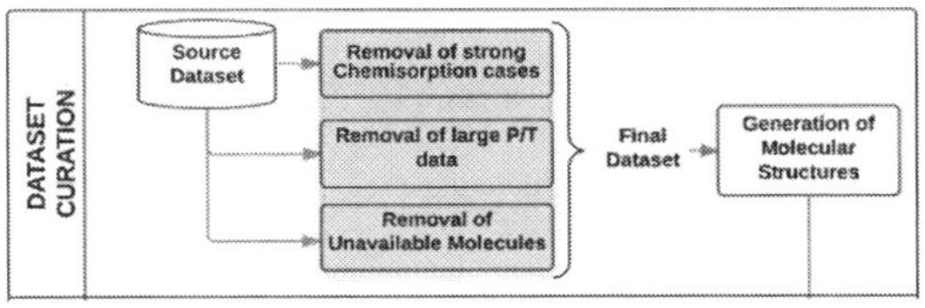

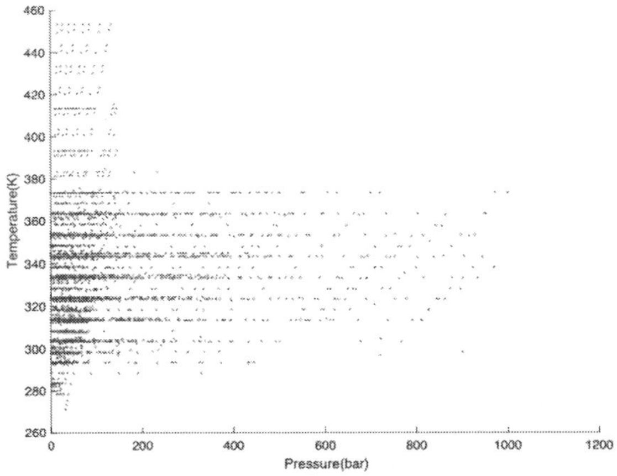

7

DFT Calculations

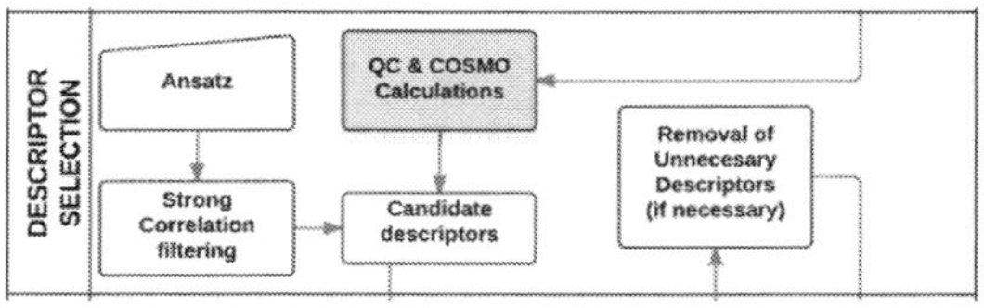

- Followed recipe in reference [1]
- C+A approach
- B3LYP functional/6-31++G** basis set
- Calculations done with TURBOMOLE
- Perfect conductor solvation
- COSMO cavity construction: Delley

Palomar et al. Understanding the physical absorption of CO2 in ionic liquids using COSMO-RS method. Industrial & Engineering Chemistry Research 2011, 50(6), 3452-346

COSMO-RS

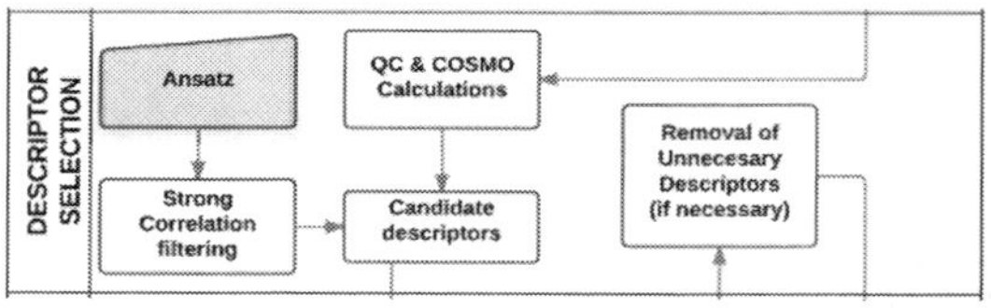

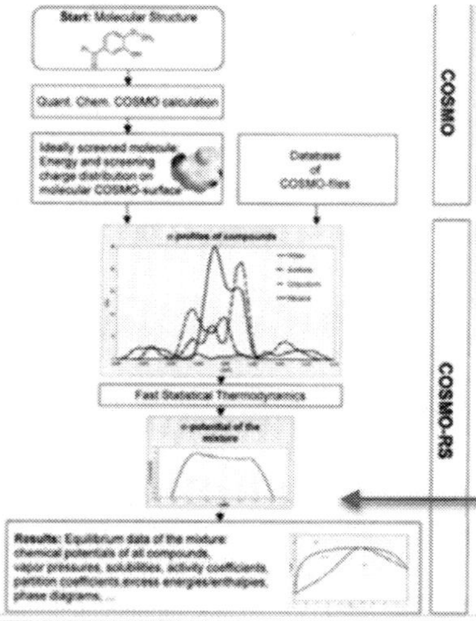

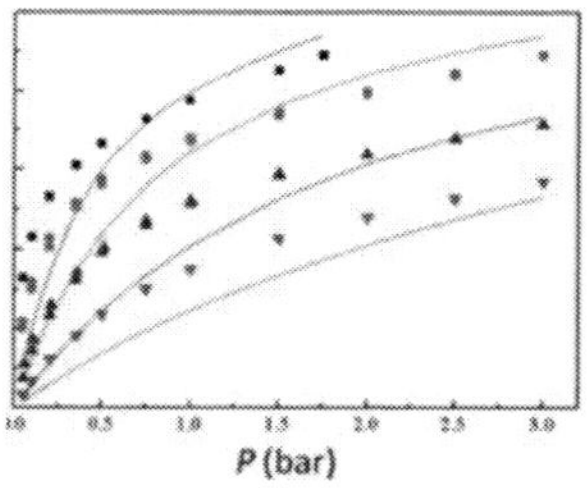

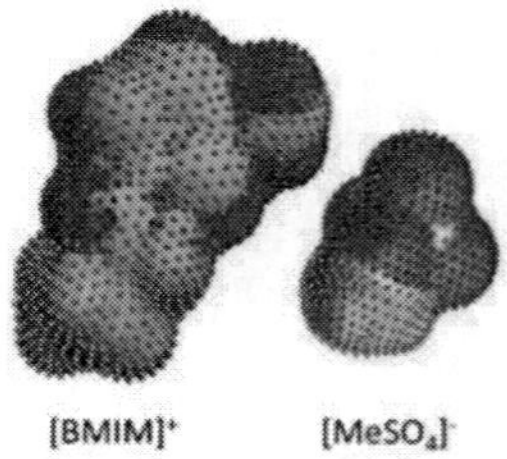

$$\mu_S(\sigma) = kT\ln\gamma_S(\sigma) = -kT\ln\left(\sum_\nu \Theta_S(\sigma_\nu)\gamma_S(\sigma_\nu)\tau(\sigma,\sigma_\nu)\right)$$
$$= -kT\ln\left(\int p_S(\sigma')\exp\left\{\frac{a_{eff}(\mu_S(\sigma')-e(\sigma,\sigma'))}{kT}\right\}d\sigma'\right)$$

[1]A. Klamt, COSMO-RS: From Quantum Chemistry to Fluid Phase Thermodynamics and Drug Design, 2005

NTNU

σ-moments

σ-moments (Chapter 9 ref[1])

$$\mu_n^-(r) = \int_{-\infty}^{r} (r-x)^n f(x)\,dx,$$
$$\mu_n^+(r) = \int_{r}^{\infty} (x-r)^n f(x)\,dx.$$

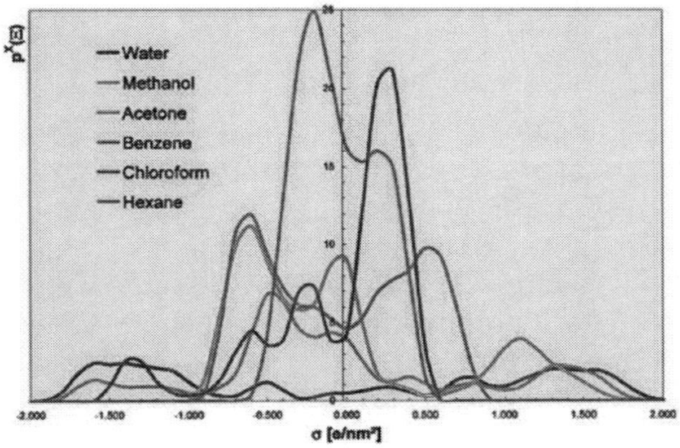

[2]Zissimoset al, A Comparison between the Two General Sets of Linear Free Energy Descriptors of Abraham and Klamt, J. Chem. Inf. Comput. Sci. 2002

10

NTNU

Filtering of descriptors

- 22 descriptors on .mom file
- 5 are energies; not used
- Surface area are obviously needed
- Correlation used to keep only one out of highly linearly related descriptors
- We ended up with the Klamt set
- 8 descriptors +(P,T)

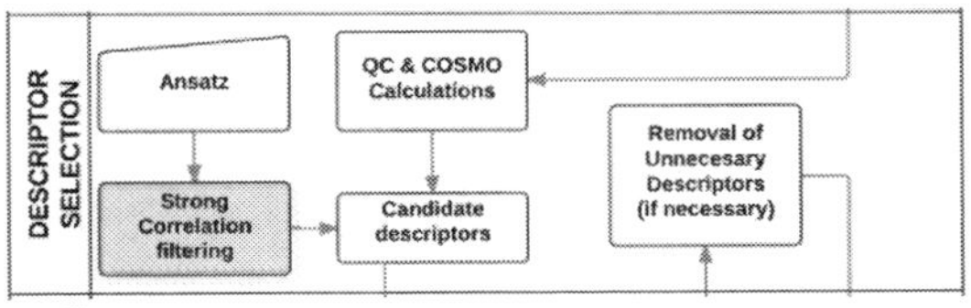

σ_0(Area)
Volume
σ_1(ion charge)
σ_2(polarity)
σ_3(skewness)
σ_4
σ_5
σ_6
MW
mu_gas

Hb_{acc1}
Hb_{acc2}
Hb_{acc3}
Hb_{acc4}
E_diel
av_dE
E_vdw
E_ring

Hb_{don1}
Hb_{don2}
Hb_{don3}
Hb_{don4}

11

ANN Design

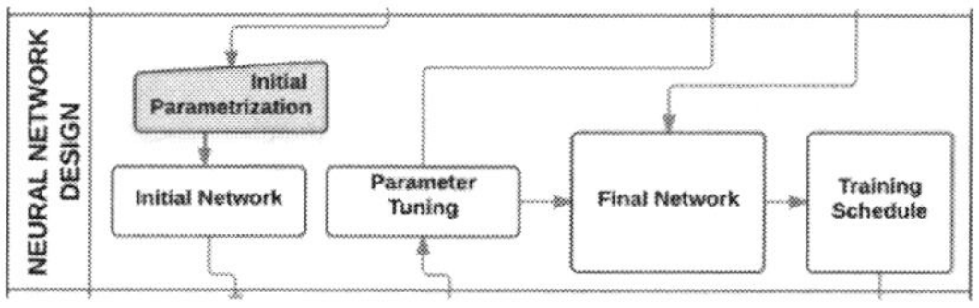

Topology	[10 10] feedforward
Layer Transfer Function	tan-sigmoidal(hidden) log-sigmoidal(output)
Loss (performance) function	Custom MSE to give each IL same weight
Back propagation algorithm	Levenberg-Marquardt
Normalization and Centering	Input layer

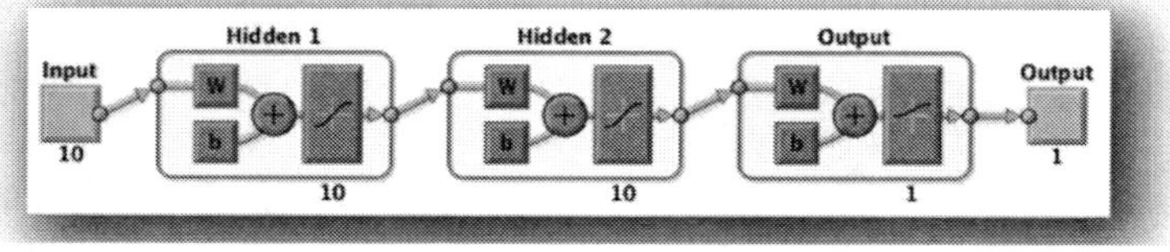

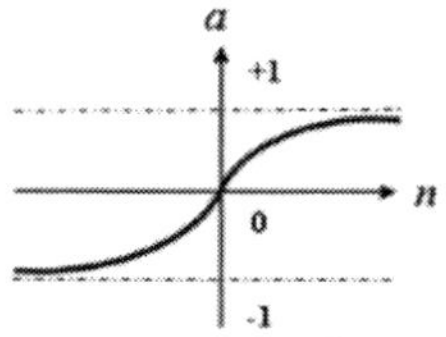

$$a = \text{Tansig}\,(n) = \frac{2}{1+e^{-2n}} - 1$$

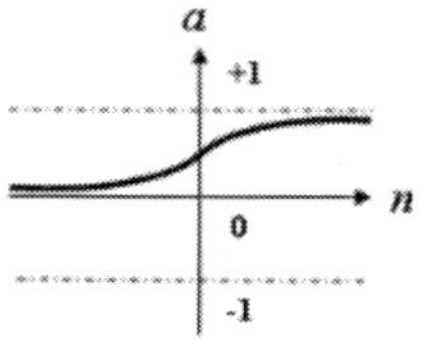

$$a = \text{Logsig}\,(n) = \frac{1}{1+e^{-n}}$$

12

Preventing overtraining, overfitting and achieving generalization

- Approaches:
  - Early Stopping

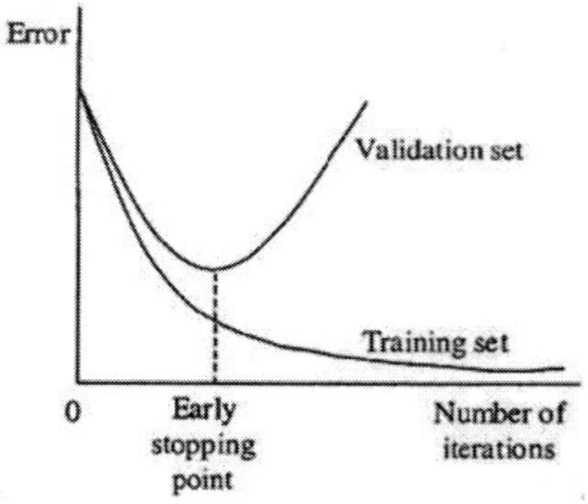

13

Preventing overtraining, overfitting and achieving generalization

- Approaches:
  - Early Stopping
  - Regularization

~~L_1-regularization~~
L_2-regularization
~~Bayesian regularization~~

13

Preventing overtraining, overfitting and achieving generalization

- Approaches:
  - Early Stopping
  - Regularization
  - y-randomization

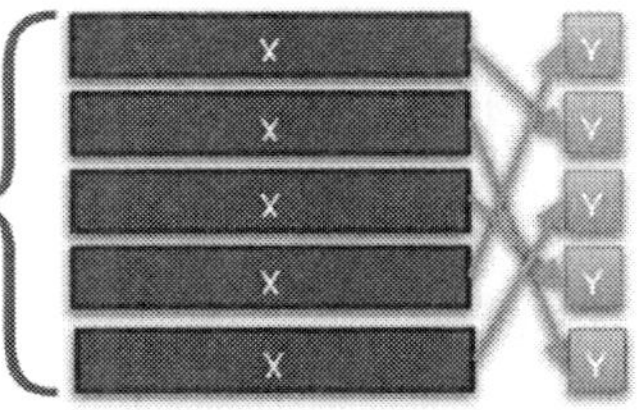

Over data points: meaningless
Over Ionic Liquids: easy to pass
Over families of ions: not well defined

13

Preventing overtraining, overfitting and achieving generalization

- Approaches:
  - Early Stopping
  - Regularization
  - y-randomization
  - Leave-one-out cross-validation

NTNU 13

Preventing overtraining, overfitting and achieving generalization

- Approaches:
  - Early Stopping
  - Regularization
  - y-randomization
  - Leave-one-out cross-validation
  - Bootstrapping

NTNU 13

NTNU

Preventing overtraining, overfitting and achieving generalization

- Approaches:
  - Early Stopping
  - Regularization
  - y-randomization
  - Leave-one-out cross-validation
  - Bootstrapping
- and of course, we first extract an external test set for final performance determination

13

NTNU

Applicability Domain Analysis (ADAN)

- ADAN defines 6 reliabilities criteria on a auxiliary PLS space of the descriptors
  d_1: distance to model
  d_2: distance closest IL in training set
  d_3: distance to centroid of training set
  d_4: difference between the predicted x and average in training set
  d_5: difference between predicted x and observed for closest IL in training set
  d_6: SDEP of the 5% of closest compounds in training set to the query compound
- d_i is violated > 95% of training set
- #violations<=2: inside AD
- #violations<=4: unreliable
- #violations>4: outside AD

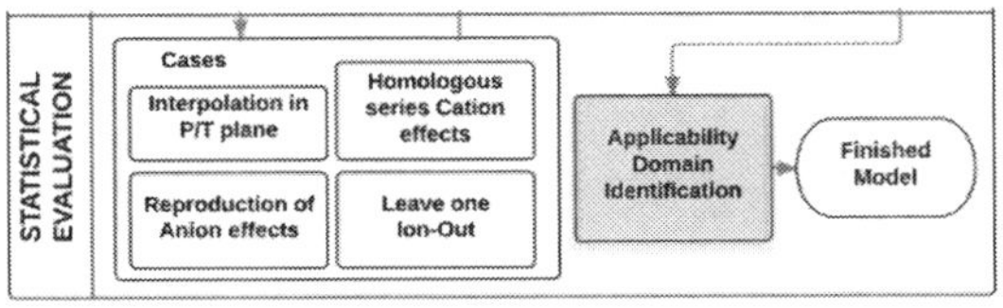

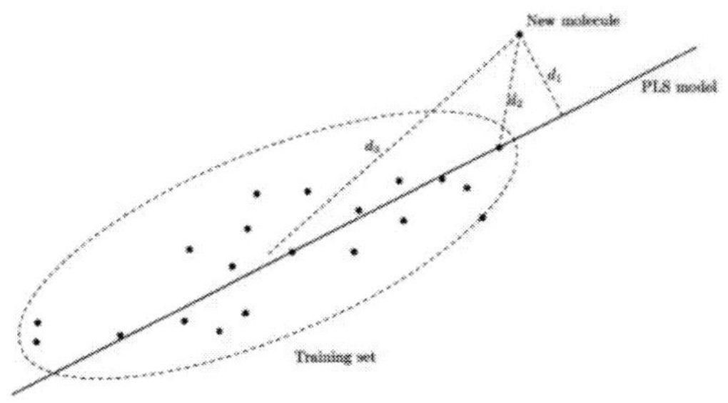

Carrio, P. et. al. *J. Chem. Inf. Mod*, 54,1500-1511 (2014)

14

Applicability Domain Analysis (ADAN)

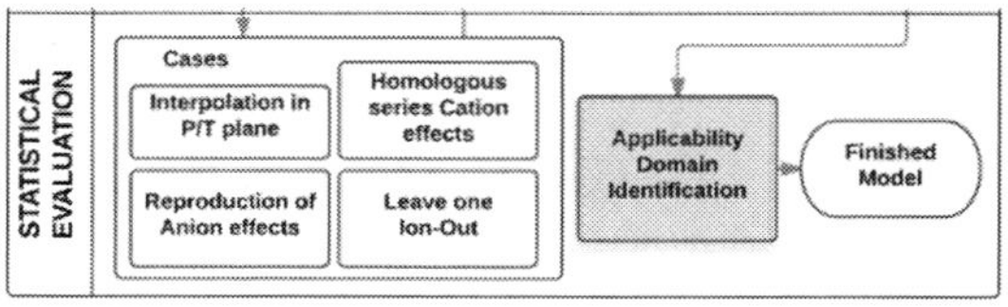

- ADAN defines 6 reliabilities criteria on a auxiliary PLS space of the descriptors
  d_1: distance to model
  d_2: distance closest IL in training set
  d_3: distance to centroid of training set
  d_4: difference between the predicted x and average in training set
  d_5: difference between predicted x and observed for closest IL in training set
  d_6: SDEP of the 5% of closest compounds in training set to the query compound
- d_i is violated > 95% of training set
- #violations<=2: inside AD
- #violations<=4: unreliable
- #violations>4: outside AD

Table 6: ADAN Violations in the Cation series effect test.

d_1	d_2	d_3	d_4	d_5	d_6	frequency
✓	✓	✓	✓	✓	✓	3827
✓	✓	✓	✓	✓	×	208
✓	✓	✓	✓	×	✓	9
✓	✓	✓	✓	×	×	2
✓	✓	✓	×	✓	✓	117
✓	✓	×	✓	✓	✓	201
✓	✓	×	✓	✓	×	10
✓	✓	×	✓	×	✓	3
✓	✓	×	×	✓	✓	2
×	✓	✓	✓	✓	✓	130
×	✓	✓	×	✓	✓	85
×	✓	×	✓	✓	✓	12
×	✓	×	×	✓	✓	3

# of Violations	count	%
0	3827	83.03
1	665	14.42
2	114	2.47
3	3	0.06

14

Performance evaluation

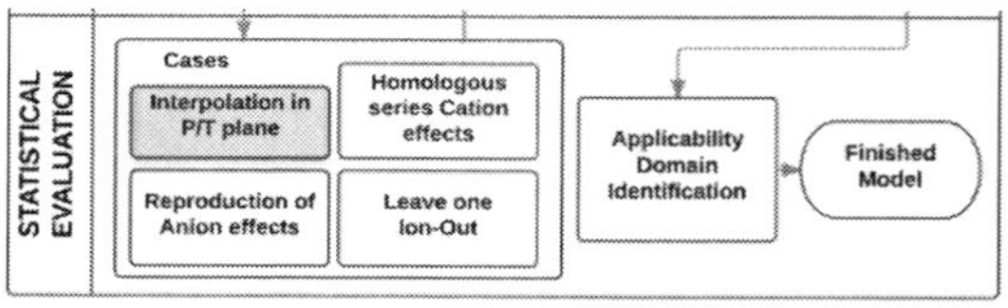

- Interpolation on known Ionic Liquids

TABLE 1. Topology parametrization (number of layers and neurons) of the neural network when working as Interpolator.

Topology		Train set		Test set		IL mean		Worst single IL	
First Layer	Second Layer	MSE	R^2	MSE	R^2	MSE	R^2	MSE	R^2
8	0	0.0009	0.992	0.0012	0.990	0.001	0.986	0.0094	0.682
10	0	0.0008	0.993	0.0010	0.992	0.001	0.984	0.0056	0.867
12	0	0.0008	0.993	0.0008	0.993	0.001	0.983	0.0053	0.707
8	8	0.0008	0.994	0.0007	0.994	0.001	0.976	0.0059	0.879
10	10	0.0007	0.994	0.0009	0.992	0.001	0.986	0.0056	0.912
12	12	0.0006	0.995	0.0006	0.995	0.001	0.980	0.0050	0.578
Final Network									
10	10								

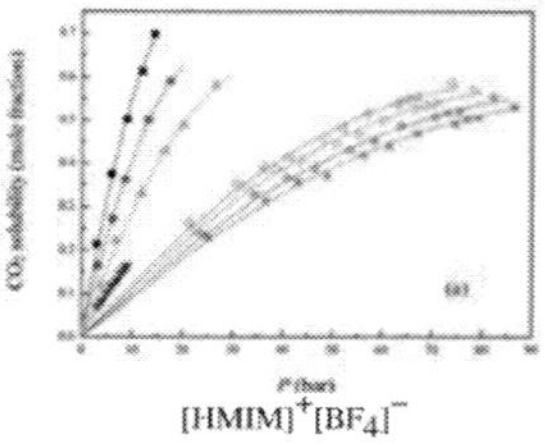

$[HMIM]^+[BF_4]^-$

ADAN: all test set points are in AD

15

Performance evaluation

- Homologous Cations

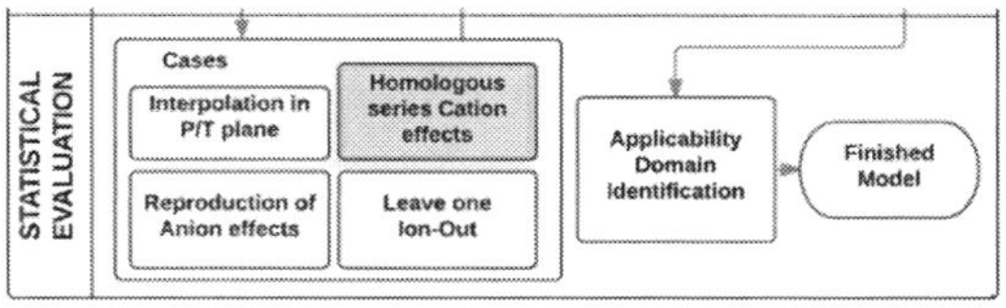

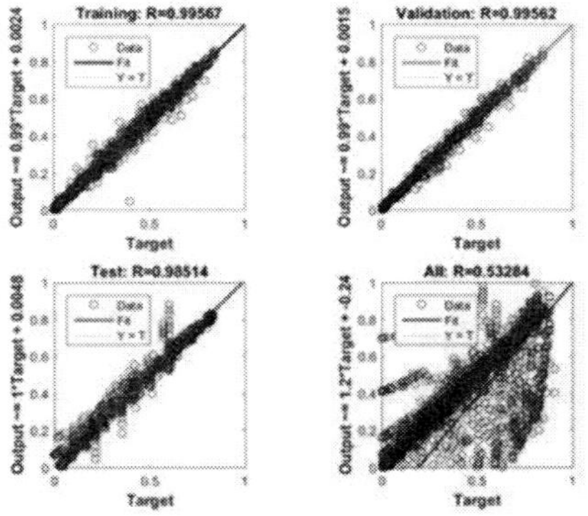

train set=[BMIM] [EMIM]
test set =[HMIM] [OMIM]

16

Performance evaluation

- The known trend is reasonably reproduced

[BF4]< [TfO]< [TfA]< [PF6]< [Tf2N]< [methide]< [C7F15CO2]< [eFAP]< [bFAP]

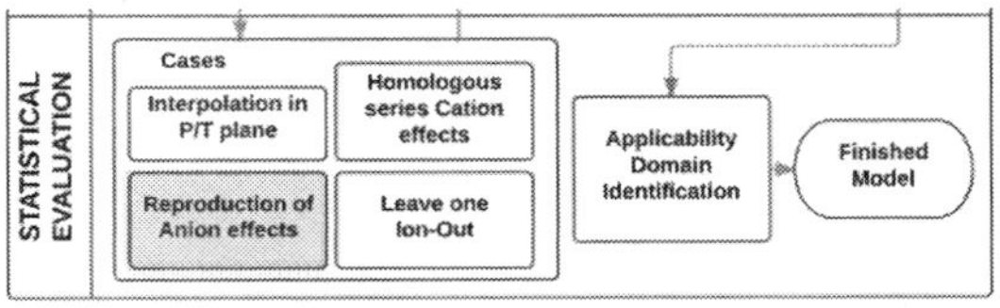

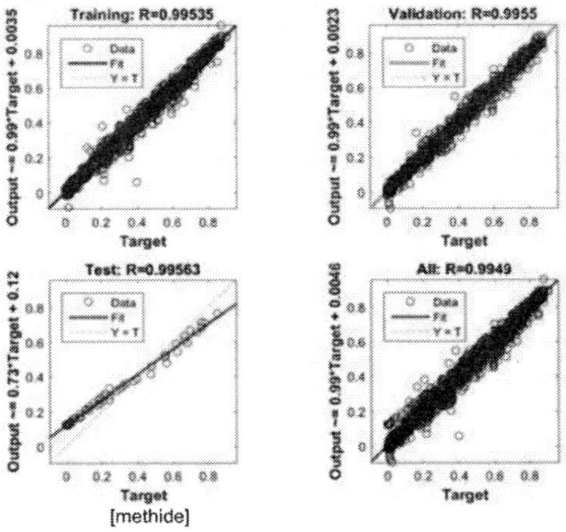

[methide]

17

Performance evaluation

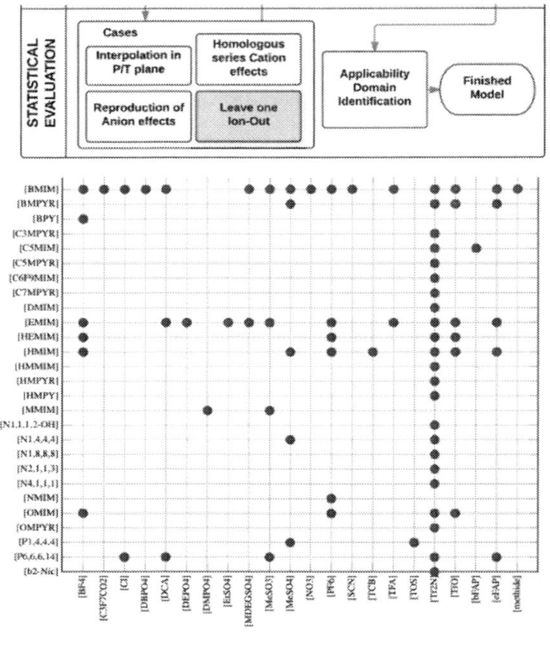

18

Performance evaluation

- Prediction when anion and cation are available in train set

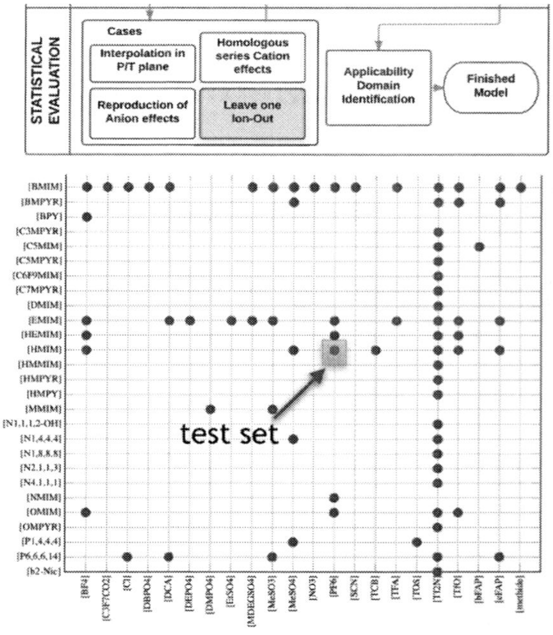

18

NTNU

Performance evaluation

- Prediction when anion and cation are available in train set
- Prediction when either Cation or Anion is unavailable

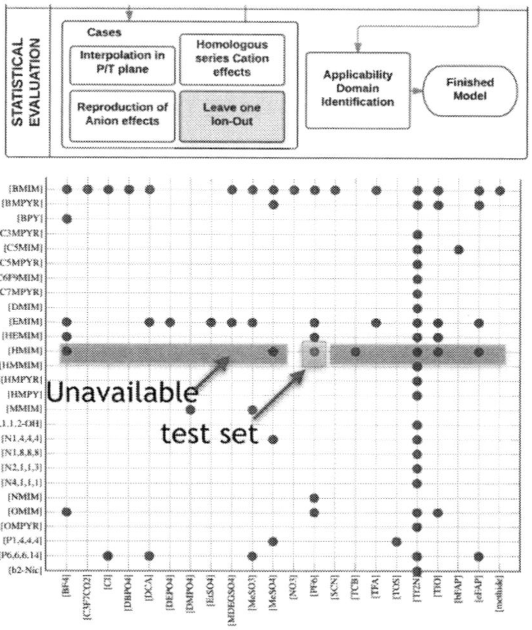

18

NTNU

Performance evaluation

- Prediction when anion and cation are available in train set
- Prediction when either Cation or Anion is unavailable
- Prediction when both Anion and Cation are unavailable

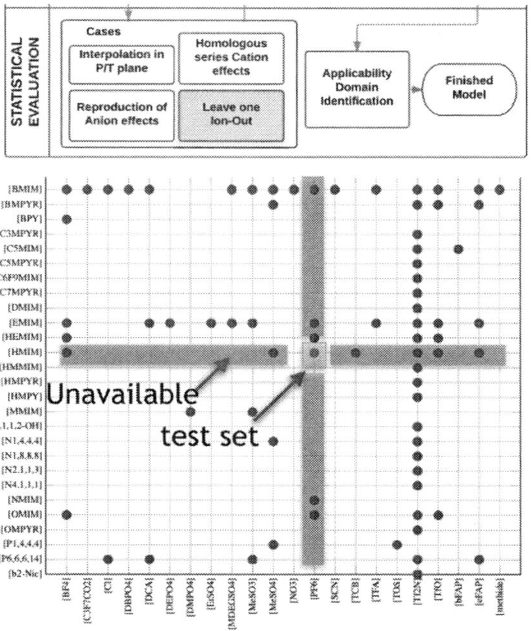

18

18

Performance evaluation

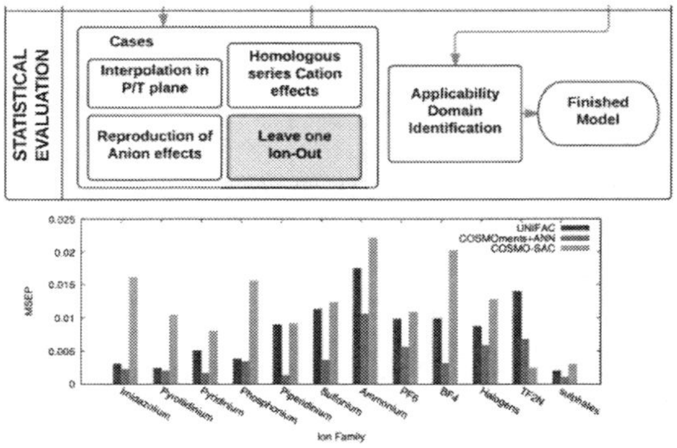

- Prediction when anion and cation are available in train set
- Prediction when either Cation or Anion is unavailable
- Prediction when both Anion and Cation are unavailable

19

Conclusions

- We presented a novel tool for the screening and prediction of CO_2 solubility in ionic liquids(ILs) on a wide range of pressures and temperatures
- Used ADAN to establish the applicability domain of the method
- The quality of prediction is competitive with COSMO-SAC and UNIFAC benchmarks presented in the review article.
- Current Wishlist:
  - Extend dataset to incorporate ILThermo
  - Mixtures of Ionic liquids & other gases: challenging because of lack of data

NTNU

Thank you for your attention

Questions?

The presenting author acknowledges the Research Council of Norway NFR project n°209337 and The Faculty of Natural Science and Technology, Norwegian University of Science and Technology (NTNU) for financial support.

20

470730 A Molecular Dynamics Investigation of Enhanced Cellulose Solubility in Ionic Liquids with Added Cosolvents

Abstract:

A number of ionic liquids (ILs) have proven to be exceptional solvents of cellulose with significant advantages over today's conventional solvents. High material costs, energy intensive solvent recovery and limited mass transport in the resultant solutions, however, have slowed their adoption in industrial processes. It has been shown that the addition of cosolvents such as DMSO and DMF, to name just a few, can substantially alleviate these mass transport issues while reducing the amount of ionic liquid that is required for dissolution. In some cases, it has been shown that adding cosolvents not only modifies the rheology of solutions, allowing dissolution to take place faster, but that it can actually enhance solubility. Furthermore, for ILs with small amounts of dissolved water, the addition of these same cosolvents can significantly improve solubility. The precise mechanisms by which these cosolvents interact with the other components of the system, leading to this solubility enhancement still remains unclear. In this study explicit, all-atom molecular dynamics simulations are used to study a select number of points on experimentally determined liquid-solid phase diagram. Simulations are performed for both the crystalline and dissolved states of cellulose with different solvent proportions and the thermodynamics of decrystallization calculated. Entropy changes are calculated using the Two-Phase Thermodynamic method allowing us to calculate the Gibb's free energy of dissolution. We find that the calculated changes agree quite well with the experimental phase diagram. Together with an analysis of hydrogen bonding changes and the three dimensional solvent structure, we provide a detailed physical picture for how the addition of moderate amounts of cosolvent enhance cellulose solubility in neat and aqueous ionic liquids.

471391 Molecular Dynamics Simulation of Binary and Ternary Deep Eutectic Solvents

Abstract:

Solvent use is of great importance to chemical and biological industries. Solvents are crucial to several processing steps, including reaction, separation, and stabilization. In several instances, the types of conventional organic solvents used are both potentially toxic to an individual, as well as to the environment. As an alternative, deep eutectic solvents (DES) are biologically based ionic liquid eutectic mixtures which possess tunable properties according to the selected cation, anion, and hydrogen bond donating co-solvent partner. These tunable properties include density, dielectric constant, and solvating power. DES possess extremely low vapor pressures, and therefore do not volatilize as easily as conventional organic solvents used in the pharmaceutical industry. Moreover, because DES are comprised of biologically based ions (i.e. choline) rather than synthetic ions (i.e. 1-ethyl-3-methyl-imidazolium), they are less toxic to organisms and the environment, and less expensive.

This work seeks to characterize environmentally benign, non-toxic, tunable DES through a combination of experimentation and molecular modeling and simulation. In particular, molecular dynamics (MD) simulations are used to compute densities, coefficients of thermal expansion, and Hildebrand solubility parameters for various binary and ternary DES, comprised of mixtures of choline chloride, urea, and glycerol. Simulation predictions of density compare well to experimental measurements. In addition, radial distribution functions (RDF) are computed to interpret the local fluid structure. RDF analysis highlights several interesting comparisons (based on composition) between urea and glycerol as hydrogen bond donors (interacting with choline and chloride) within DES, as well as unique hydrogen bonding between urea and glycerol molecules within the DES mixture.

454116 Self-Assembly Induced Solubilization of Drug-like Molecules in Nanostructured Ionic Liquids

Abstract:

Dissolution is a fundamental process that is vital for chemistry and biological science. Unfortunately, many drug molecules exhibit poor solubility due to their high molecular weights and structural complexity, which leads to low bioavailability and substantially hinders the development of efficient separation methods and drug delivery systems. To date, a wide variety of solubilization strategies have been developed to improve drug molecule solubility. These include molecular modification, mixed solvents, micelles and microemulsions, cosolvent, deep eutectic solvents, and ionic liquids (ILs). Although substantial progress has been made, the solubility of relatively large molecules with both H-bond and hydrophobic segments is still limited. Aim of the present work is to overcome this limitation, a novel and versatile approach, which is referred to as a self-assembly induced (SAI) solubilization strategy is developed using nanostructured ILs, for the dissolution of sparingly soluble drug-like molecules with both H-bond and hydrophobic segments. First, the drug-like molecules form an amphiphilic complex with the nanostructured ILs *via* H-bond interactions, and then, the complex undergoes secondary self-assembly in the nanostructured ILs to form a highly ordered aggregate structure, such as a liquid crystal (LC), at room temperature. With this strategy, the enhancement of solubilization could transcend the limit of the conventional interaction-adjustment mode, resulting in extremely high solubilities of various drug-like molecules being achieved. For example, the molar solubilities of cholesterol and stigmasterol were as high as 0.91 and 0.93 at 50 °C, which are 5-8000-fold higher than their solubilities in organic solvents, micelles, microemulsions, and common ILs; the solubilities of naproxen, folic acid, hydrocortisone, vitamin D_3, and indomethacin, are also the highest level reported in the literature. Small angle X-ray scattering (SAXS), polarized optical microscopy (POM), wide angle X-ray diffraction (WAXRD) and infrared (IR) spectroscopy measurements were employed to investigate the nanoscale organizations in these long-chain carboxylate ILs (LCC-ILs) and the dissolution mechanism. In addition to the ultra-high solubility, the SAI solubilization strategy using nanostructured ILs has two distinctive features. First, differing from the classic microemulsion solubilization method that uses surfactants, solvents, and other excipients, only one LCC-IL was employed in this strategy due to the unique properties of these nanostructured ILs. Secondly, the SAI solubilization strategy is highly reversible. It is a physical dissolution process based on reversible H-bond self-assembly rather than covalent modification. In addition, the solubilities of these seven drug molecules in water/$[P_{4444}][C_{15}H_{31}COO]$ and phosphate-buffered saline (PBS)/$[P_{4444}][C_{15}H_{31}COO]$ mixtures are still 2-35-fold higher than in common microemulsions under similar conditions and is 2-4 orders of magnitude higher than in pure water even at low IL concentration. Our work shows that both mesoscopic aggregation and H-bond interactions play pivotal roles in the self-assembly induced dissolution behavior. Based on its excellent performance and unique chemistry, we believe that the SAI solubilization strategy will inspire more applications of nanostructured ILs for advanced separation and novel drug delivery systems.

References

1. Jin W. B.; Yang Q. W.; Bao Z. B.; Zhang Z. G.; Ren Q. L.; Yang Y. W.; Xing H. B. *Chem. Commun.*, **2015**, *51*, 13170-13173.
2. Jin W. B.; Yang Q. W.; Huang B. B.; Bao Z. B.; Zhang Z. G.; Ren Q. L.; Yang Y. W.; Xing H. B. *Green Chem.*, **2016**, DOI: 10.1039/C6GC00584E.
3. Yang Q. W.; Xu D.; Zhang J. Z.; Zhu Y. M.; Zhang Z. G.; Qian C.; Ren Q. L.;Xing H. B. *ACS Sustainable Chem. Eng.* **2015**, *3*, 309-316.

Acknowledgements

The research was supported by the National Natural Science Foundation of China (21222601, 21476192, and 21436010), the Zhejiang Provincial Natural Science Foundation of China (LR13B060001), and Huabin Xing was supported by the Young Top-Notch Talent of Ten Thousand Talent Program of China..

462279 Simulations of Li^+ in Ionic Liquids: Structure, Transport, and Electrochemical Windows

Abstract:

Ionic liquids have been proposed as candidate electrolytes for a number of electrochemical applications. The Li^+ solvation structure in these liquids is of central importance to electrolyte properties, like ionic conductivity and electrochemical stability. To this point, we employ simulations at three different size scales to better understand various aspects of the interplay between Li^+ solvation structure and dynamics. The smallest systems are $Li(Anion)_n$ clusters that are treated with high-accuracy density functional theory (DFT) techniques to provide insight into solvation shell structure through energetics and comparisons to experimental IR/Raman spectra. Mid-range sized liquid-phase systems (12-24 ion pairs) are treated with DFT molecular dynamics (MD) to provide temperature-dependent insight into Li^+ solvation structure, diffusion, and electrochemical window. The largest systems (144-216 ion pairs) are treated with polarizable MD simulations to evaluate the influence of Li-networks on structure and provide size independent values of transport properties. We perform this procedure on three technologically important ionic liquids and comment on property correlations with solvation structure.

475491 R&D Progress in Air Pollution Control Technologies in China (Featured talk)

Abstract:

For recent years, people in most northern and eastern parts of China have suffered from the heavily-polluted haze. Besides natural climate variations, emissions from transportation and industrial sources are the main contributors. In this presentation, current situation of industrial air pollution emissions in China will be summarized and applications of pollution control technologies will be critically reviewed with emphasis on the catalytic oxidation of volatile organic compounds (VOCs) and potential opportunities for novel materials.

461082 Removal of Toxic Chemicals Using Metal-Organic Frameworks

Abstract:

Metal-organic frameworks (MOFs) have been widely investigated for a wide range of applications, most notably gas storage, separations, and sensing. Only recently have advances been made in the area of toxic chemical remediation focusing on potentially catalytic chemistries for removal. Here we discuss the use of several metal-organic frameworks to remove chemical warfare agent as well as other less-toxic, yet still problematic, chemicals such as chlorine and alkyl halides. In particular, we focus on multiple removal mechanisms, including hydrolysis as well as nucleophilic attack, to mitigate these chemicals.

469056 Adsorption Energies of Light Hydrocarbons in Functionalized Uio-66(Zr)

Abstract:

Differential adsorption energies of ethane, ethylene, and acetylene on the metal-organic framework (MOF) UiO-66(Zr) and UiO-66(Zr) functionalized with â€"NH_2 and â€"NO_2 were measured by adsorption calorimetry at 293K using a Tian-Calvet calorimeter. Measurements were taken at low coverage ranging from 0.05 mmol/g to 1.2 mmol/g. In all cases the MOFs were heated under vacuum to dehydroxylate the MOFs prior to testing. The adsorption energy of ethylene was highest for UiO-66(Zr)-NO_2 and was lowest for UiO-66(Zr). For UiO-66(Zr)â€"NO_2, the adsorption energy of ethane was higher than ethylene and only slightly lower than acetylene. Given that the overall molecular polarity of ethane is lower than ethylene and that the kinetic diameter of ethane is larger than ethylene, confinement effects may be a dominating factor for ethane adsorption in the -NO_2 case. For adsorption of both ethane and acetylene on UiO-66(Zr)-NO_2, the adsorption energies decreased by ~5 kJ/mol as the coverage increased from 0.05 mmol/g to ~0.6 mmol/g. For ethylene, the adsorption energies were essentially constant over the entire range of coverage investigated. The change in adsorption energy for ethane and ethylene at low coverage may indicate that UiO-66(Zr)-NO_2 has multiple types of adsorption sites. We are currently investigating low-temperature calorimetry measurements for this system, which will allow one to more clearly isolate the energies of stronger and weaker adsorption sites.

473309 Selective Gas Adsorption on Functionalized H3btc and Di-Isophthalate Based Metal Organic Frameworks

Selective Gas Adsorption on Functionalized H_3BTC and Di-Isophthalate Based Metal Organic Frameworks

Prudhviraj Medikonda[1], Sastri Chivukula[2] and Sasidhar Gumma[1*]

1 Department of Chemical Engineering, Indian Institute of Technology Guwahati, India

2 Department of Chemistry, Indian Institute of Technology Guwahati, India

E-mail: *s.gumma@iitg.ernet.in*

Functionalization of organic linkers towards metal organic frameworks (MOFs) containing open metal sites having more attention and interest provides predictable pore size, surface area and more adsorption sites (Collins, 2009). The Cu-BTC MOF is a widely studied adsorbent for gases separation and storage, but limited work has been done to modify the H_3BTC ligand or functionalize Cu-BTC to optimize its properties for gas adsorption. On the other hand, organic linkers containing double-bond functionalities have rarely been studied for gas adsorption using MOFs and N containing MOFs are more selective for CO_2 over N_2, CH_4 and CO (Liu et al. 2007, Belof et al. 2007). In this work we report the effect of functionalization by comparing the gas adsorption properties of 1 with 2 and those of 3 with 4. The role of the functional group and the polarity of the gas will be discussed.

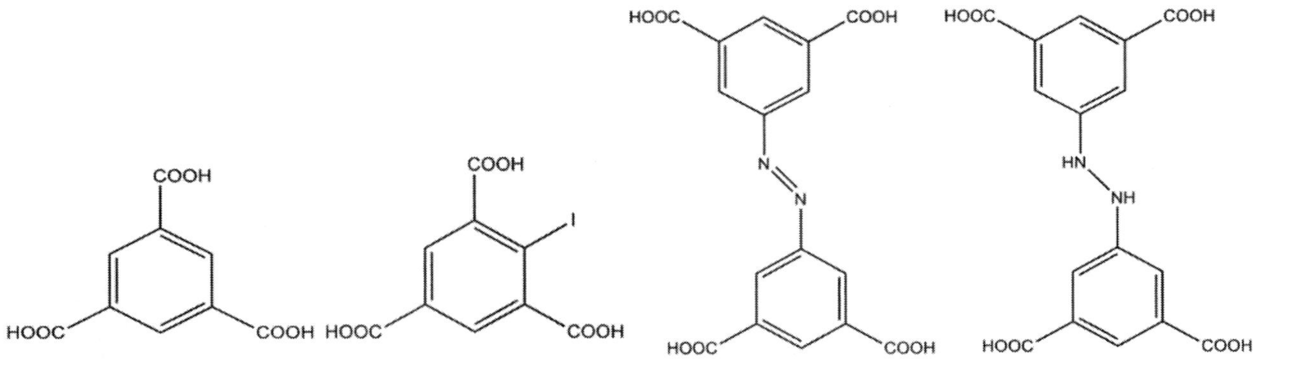

1 2 3 4

Scheme 1: Structures of organic linkers H_3BTC (**1**), H_3IBTC (**2**), $H_4AzoBTC$ (**3**) and H_4 reduced AzoBTC (**4**) used in this work

References

D. J. Collins, S. Ma, H. C. Zhou. *Hydrogen and methane storage in Metal-Organic Frameworks: Design and Application,* Wiley-VCH, Weinheim 2009.

Y. Liu, J. F. Eubank, A. J. Cairns, J. Eckert, V. Ch. Kravtsov, R. Luebke, M. Eddaoudi. *Angew. Chem. Int. Ed. 46*, 3278, 2007.

J. L. Belof, A. C. Stern, M. Eddaoudi, B. Space. *J. Am. Chem. Soc., 129*, 15202, 2007.

Y. Cai, Y. Zhang, Y. Huang, S. R. Marder, K. S. Walton. *Cryst. Growth Des.,* 12, 3709–3713, 2012.

454249 A One-Step Ion Exchange Method to Functionalized Zif-67 for Enhanced CO2 Capture

Abstract:

A one-step ion exchange method to functionalized ZiF-67 for enhanced CO_2 capture

Fujiao Song, Qin Zhong*

School of Chemical Engineering, Nanjing University of Science and Technology, Nanjing 210094, PR China. E-mail: Song_fj2006@126.com (Fujiao Song), zq304@mail.njust.edu.cn (Qin Zhong).

ZiFs have shown great potential applications in gas separation [1]. Ion exchange could improve the surface area, pore volume as well as surface charge of the adsorbents, further enhancing the CO_2 adsorption capacities. However, it is notable that the standard ion-exchange process needs to be repeated for several times, which brings a series of disadvantages, such as complicated in synthesis, time consuming and solvent consuming [2, 3].

In this work, we develop a new one-step ion exchange method (donated as ion as-exchange technique) to overcome the disadvantages mentioned above. Since the cations for exchange (Li^+ and Na^+) was added before the hydrothermal preparation of ZiF-67, time consuming and solvent consuming in the standard ion-exchange procedure was eliminated. The product was designated as M-as, where ¡°M¡± represents the alkali metal ion (Li and Na). By comparison, the standard ion-exchange procedure was also carried out to synthesize the adsorbents, and the product was designated as M-post. The structure, surface charge and CO_2 adsorption properties of the adsorbents prepared via the two methods were investigated and compared.

Fig. 1(A) shows the XRD patterns of ZiF-67 exchanged with Li^+ and Na^+. all the ion exchanged ZiF-67 maintains crystallinity of the original phase [4]. And almost no changes in peak intensity and position could be observed. Fig. 1(B) presents the particle size distributions of as-synthesized and ion exchanged ZiF-67 samples, which shows the Gaussian type distribution with different distribution ranges.

The CO_2 adsorption isotherms of parent and these ion-exchanged ZiFs at 0 ¡ãC are shown in Fig. 1(C). The CO_2 uptake of parent ZiF-67 at 1 bar is 35.4 cm^3/g. Li/Na-post exhibited decreased CO_2 adsorption capacity (30.4 and 31.2 cm^3/g) compared to the parent, while Li/Na-as showed improved CO_2 uptake values (48.7 and 45.1 cm^3/g). The highest CO_2 capacity is achieved by Li-as, which is 37.6 % higher than that of the parent ZiF-67. An interesting phenomenon is found that all of the isotherms show linear patterns with stable adsorption rate from 0 bar to 1 bar, which reveals the materials could get excellent adsorption performance at higher pressure range (> 1 bar).

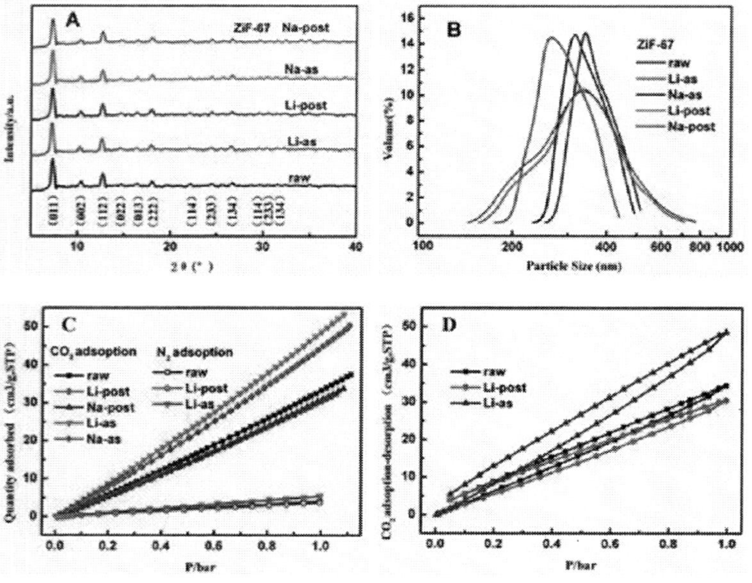

Fig. 1. (A) XRD patterns of parent ZiF-67 and ion exchanged with Li^+ and Na^+, (B) Particle size distribution, (C) CO_2/N_2 adsorption isotherms and (D) CO_2 adsorption-desorption isotherms

The CO_2 adsorption mechanism of the adsorbents prepared via the two methods was investigated. The surface area and pore volume (Table 1) of Li^+/Na^+ as-exchanged ZiF-67 increase significantly because the radius of Li^+ and Na^+ is smaller than that of Co^{2+}. However, the surface area of post-exchanged materials decreases contrarily due to the agglomerate and pore block. PZC characterization (Table 2) indicates that parent ZiF-67 has a neutral character, whereas ion-exchanged ZiFs has basic surfaces. And pH_{PZC} values of Li^+ and Na^+ as-exchanged ZiFs is higher than that of post-exchanged samples, which could improve CO_2 adsorption due to the coordination interaction. All of the CO_2 adsorption isotherms show linear patterns with stable adsorption rate from 0 bar to 1 bar, which reveals the materials could get excellent adsorption performance at higher pressure range (> 1 bar). Van der Waals interaction determined by the surface area and coordination interaction resulting from electrostatic interaction [5] work in synergy to enhance CO_2 adsorption performance of Li^+/Na^+ as-exchanged ZiF-67 adsorbents.

Table 1. Physical properties of parent ZiF-67 and ion exchanged ZiF-67.

ZiFs	BET (m^2/g)	Pore volume (m^3/g)	SF Median pore width (nm)
parent	1424.59	0.84	0.868
Li-as	1450.71	0.87	0.875
Na-as	1435.15	0.86	0.871
Li-post	1366.14	0.76	0.851
Na-post	1359.50	0.77	0.864

Table 2. The pH_{PZC} values of the parent ZiF-67 and ion exchanged ZiF-67 before and after CO_2 adsorption

Sample	pH_{PZC} values	
	before CO_2 adsorption	after CO_2 adsorption
parent	6.92	5.36
Li-post	7.79	5.10
Na-post	7.51	5.63
Li-as	9.26	5.27
Na-as	8.83	5.49

References

[1] J. An, N.L. Rosi, J. Am. Chem. Soc. 132 (2010) 5578-5579.

[2] M. Eddaoudi, J. Eubank, Y. Liu, V.Ch. Kravtsov, R. Larsen, J. Brant, Stud. Surf. Sci. Catal. 170 (2007) 2021-2029.

[3] F. Nouar, J. Eckert, J. Eubank, P. Forster, M. Eddaoudi, J. Am. Chem. Soc. 131 (2009) 2864- 2870.

[4] J. Qian, F. Sun, L. Qin, Mater. Lett. 82 (2012) 220-223.

3601-3606.

[5] H. Zhuo, Q. Li, W. Li, J. Cheng, New J. Chem. 39 (2015) 2067-2074.

*Corresponding author. Tel.:+86 025-84315517; fax: +86 025-84315517.

E-mail: zq304@mail.njust.edu.cn (Qin Zhong).

453025 CO_2 Capture in Uio-66-Type Metal-Organic Frameworks (MOFs)

Abstract:

CO_2 Capture in UiO-66-Type Metal-Organic Frameworks (MOFs)

Dan Zhao*

Department of Chemical and Biomolecular Engineering,

National University of Singapore, 117585 Singapore

E-mail: chezhao@nus.edu.sg

The large-scale anthropogenic carbon dioxide (CO_2) emission has aroused world-wild concerns. One of the key steps in carbon mitigation is CO_2 capture from stationary CO_2 emission resources. Metal-organic frameworks (MOFs) are coordination polymers consisting of metal ions or clusters as nodes and organic ligands as spacers. The study of CO_2 capture using MOFs as adsorbents is booming because of their ultra-high surface area, uniform yet tunable pore size, and versatile chemical compositions. However, several issues still exist preventing their large scale applications, such as selectivity, stability, and material cost. In our research, we aim at circumventing the above problems through rational ligand design and crystal engineering based on UiO-66-type MOFs, with the emphases on a) mitigating moisture interference, b) increasing CO_2 selectivity, c) strengthening materials stability, and d) reducing material cost. The synthesized MOF candidates have underwent various tests for their eligibility in CO_2 capture, such as stability, gas sorption, and breakthrough tests. We have identified several promising MOF candidates with the CO_2 capture performance equivalent or even better than the industrial benchmark material zeolite 13X. These materials will be integrated into the next-level process and system research.

References

1. Hu, Z. G.; Castano, I.; Wang, S. N.; Wang, Y. X.; Peng, Y. W.; Qian, Y. H.; Chi, C. L.; Wang, X. R.; Zhao, D.* Modulator effects on the water-based synthesis of Zr/Hf MOFs: quantitative relationship studies between modulator, synthetic condition, and performance. *Cryst. Growth Des.* **2016**, DOI: 10.1021/acs.cgd.6b00076.
2. Hu, Z. G.; Nalaparaju, A.; Peng, Y. W.; Jiang, J. W.; Zhao, D.* Modulated hydrothermal synthesis of UiO-66(Hf)-type metal-organic frameworks for optimal carbon dioxide separation. *Inorg. Chem.* **2016**, *55*, 1134-1141.
3. Hu, Z. G.; Faucher, S.; Zhuo, Y. Y.; Sun, Y.; Wang, S. N.; Zhao, D.* Combination of optimization and metalated-ligand exchange: an effective approach to functionalize UiO-66(Zr) MOFs for CO_2 separation. *Chem.-Eur. J.* **2015**, *21*, 17246-17255.
4. Hu, Z. G.; Peng, Y. W.; Kang, Z. X.; Qian, Y. H.; Zhao, D.* A modulated hydrothermal (MHT) approach for the facile synthesis of UiO-66-type MOFs. *Inorg. Chem.* **2015**, *54*, 4862-4868.
5. Hu, Z. G.; Khurana, M.; Seah, Y. H.; Zhang, M.; Guo, Z. G.; Zhao, D.* Ionized Zr-MOFs for highly efficient post-combustion CO_2 capture. *Chem. Eng. Sci.* **2015**, *124*, 61-69.
6. Hu, Z. G.; Zhang, K.; Zhang, M.; Guo, Z. G.; Jiang, J. W.; Zhao, D.* A combinatorial approach towards water-stable metal-organic frameworks for highly efficient carbon dioxide separation. *ChemSusChem* **2014**, *7*, 2791-2795.

461320 Tuning the Kinetic Water Stability and Adsorption Interactions of Mg-MOF-74 By Partial Substitution with Co or Ni

Abstract:

Varying amounts of Co and Ni were substituted into the metal-organic framework Mg-MOF-74 via a one-pot solvothermal reaction, and the effects of these substitutions on CO_2 adsorption and kinetic water stability properties were examined. Based on elemental analyses, Co and Ni are more favorably incorporated into the MOF-74 framework from solution than Mg. In addition, reaction temperature more strongly impacts the final metal composition in these mixed-metal (MM) MOF-74 structures than does the reaction solvent composition. Single-component CO_2 adsorption isotherms were measured for the MM-MOF-74 systems at 5, 25 and 45 °C and isosteric heats of adsorption were calculated. These results suggest that CO_2 sorption properties can be adjusted by partial metal substitution. Water adsorption isotherms were also measured for the MM-MOF-74 samples, with powder X-ray diffraction patterns and Brunauer-Emmett-Teller surface areas measured both before and after water exposure. Results show that Mg-MOF-74 can gain partial kinetic water stability by the incorporation of Ni^{2+} or Co^{2+} metal ions that are less vulnerable to hydrolysis than Mg^{2+}. Of particular note, Mg-Ni-MM-MOF-74 shows a significant increase in water stability when incorporating as little as 16 mol% Ni into the Mg-MOF-74 structure.

Energy and CO_2 Emissions Intensity of Desalination Systems

Presenter:
Caroline Dollinger
Engineer, Energy Analysis
Energetics Incorporated

American Institute of Chemical Engineers
Annual Meeting
San Francisco, CA
November 17, 2016

Collaboration

Caroline Dollinger (Speaker)
Energetics Incorporated
cdollinger@energetics.com

Sabine Brueske
Energetics Incorporated
sbrueske@energetics.com

Prakash Rao
Lawrence Berkeley National Laboratory
prao@lbl.gov

Joseph Cresko
U.S. Department of Energy
Advanced Manufacturing Office
joe.cresko@ee.doe.gov

Arian Aghajanzadeh
Lawrence Berkeley National Laboratory
aaghajanzadeh@lbl.gov

Energetics Incorporated is an energy consulting company, providing subject matter expertise to private and public sector clients

2

Overview

- An *Energy Bandwidth Study* is currently being performed to analyze energy use and energy saving opportunities in desalination
- Energy data will broken down by unit operation to identify processes and energy-saving technologies with the greatest opportunities for energy savings
- Results with also show the CO_2 emissions savings

Results presented in this talk are <u>draft data</u>; the work is still ongoing and a final report is expected to be published next year.

3

BERKELEY LAB | U.S. DEPARTMENT OF ENERGY | Energy Efficiency & Renewable Energy | ADVANCED MANUFACTURING OFFICE | ENERGETICS A Subsidiary of VSE Corporation

Energy bandwidth studies can be a useful tool for assessing energy savings opportunities in industry

Energy bandwidth studies frame the range (or *bandwidth*) of potential energy savings in industry, and technology opportunities to realize those savings.

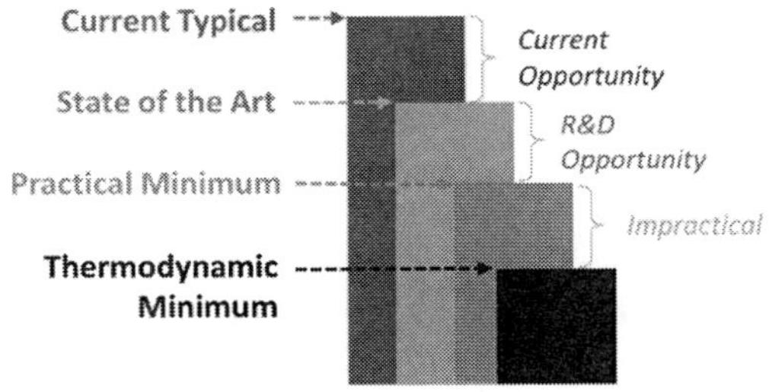

Measures of energy intensity studied:

Current Typical (CT)	State of the Art (SOA)	Practical Minimum (PM)	Thermodynamic Minimum (TM)
Determined from a literature review and stakeholder outreach, based on current typical industrial processes in the U.S.	Determined from a literature review and stakeholder outreach, based on the most energy-efficient technologies and practices available worldwide	Modeled based on plausible energy savings from identified R&D technologies under development worldwide	Calculated analytically using a Gibbs free energy approach assuming ideal conditions

4

BERKELEY LAB | U.S. DEPARTMENT OF ENERGY | Energy Efficiency & Renewable Energy | ADVANCED MANUFACTURING OFFICE | ENERGETICS A Subsidiary of VSE Corporation

Six new energy bandwidth studies on lightweight materials were recently completed in draft form

Summary thumbnail for bandwidth study on petroleum refining

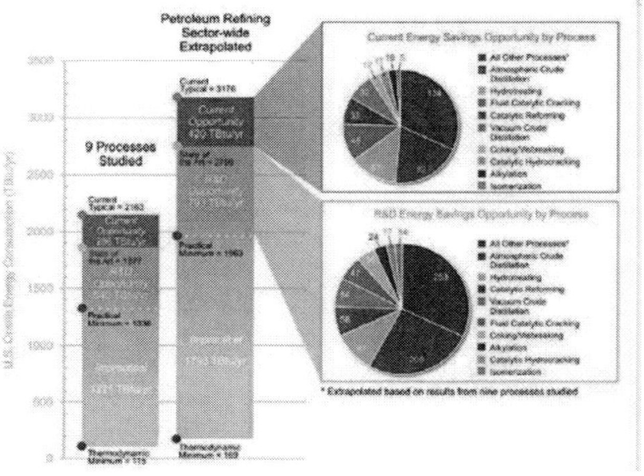

- **These studies have previously provide a deep-dive analysis into manufacturing energy use and potential energy savings for:**
  - **Four energy-intensive manufacturing sectors**
  - **Manufacturing of six lightweight materials**
- **This talk will focus primarily on a new energy-water bandwidth of desalination systems.**
- **Published bandwidths (including this one, once completed) are available on the DOE Advanced Manufacturing Office website**
  - http://energy.gov/eere/amo/energy-analysis-sector#5

5

BERKELEY LAB U.S. DEPARTMENT OF ENERGY Energy Efficiency & Renewable Energy ADVANCED MANUFACTURING OFFICE ENERGETICS A Subsidiary of VSE Corporation

First Four Industry Bandwidths – US Results

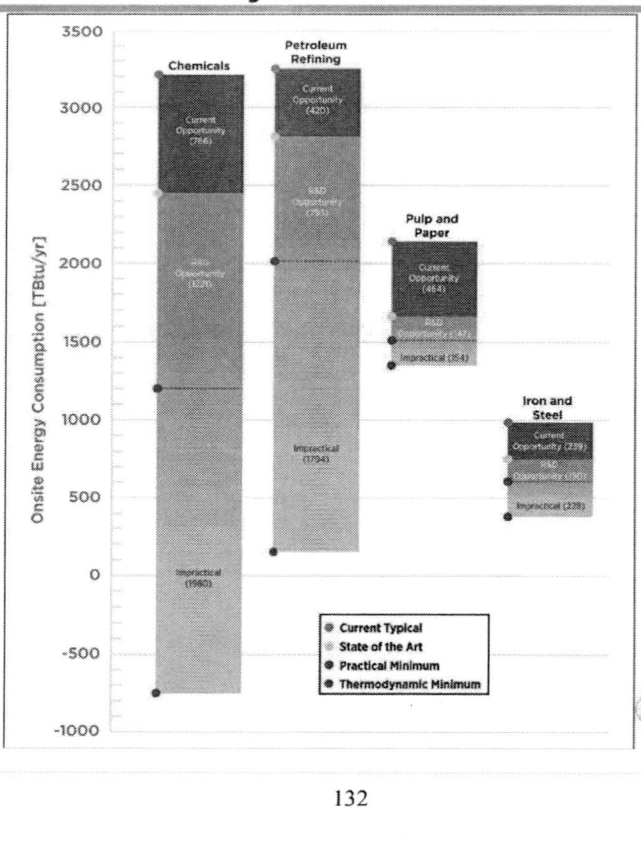

6

ENERGETICS A Subsidiary of VSE Corporation

Overview

- **Review of the current findings of Volume 1 report, literature survey**
- **Plans for Volume 2 report, bandwidth study**
  - **Includes "scenarios" for desalination uptake in the United States for the production of potable water from seawater**
- **Including discussions of:**
  - **Study scope**
  - **Unit operations studied**
  - **Boundary analysis**
  - **Technologies reviewed**
  - **Preliminary energy data**
  - **Key findings**

7

Energy Water Studies (Current and possible)

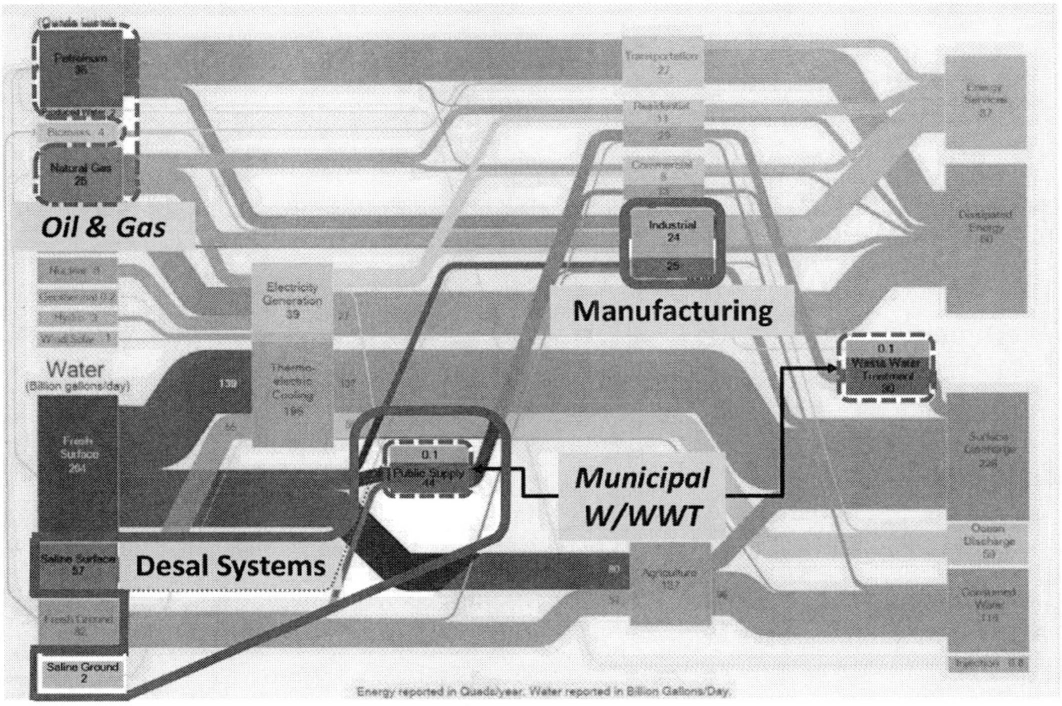

Two Volumes to Energy-Water BW Study

Volume	Contents	Status
Volume 1 – Survey of Available Information in Support of Energy-Water Bandwidth Study of Desalination Systems	• Energy intensities for five unit operations of desalination system • Boundary analysis framework • Framework for establishing desalination uptake scenarios, calculating total water cost, and carbon emissions	• Published, available on LBNL website (https://eetd.lbl.gov/publications/volume-1-survey-of-available-informat)
Volume 2 – Energy-Water Bandwidth Study of Desalination Systems	• Energy and CO_2 Consumption for Several Sea and Brackish Water to Municipal Water Scenarios Evaluated at CT, SOA, PM, TM • Current Energy and CO_2 Savings Opportunity • R&D Energy and CO_2 Savings Opportunity	• Research and analysis underway • Expected to be completed 2017

9

"Red Team" Reviewers

Industry

- Tom Pankratz: independent consultant, Editor of Water Desalination Report
- Lisa Henthorne, PE: senior VP and CTO at Water Standard
- Harold Fravel , PhD: Executive Director at American Membrane Technology Association

Government

- Frank Leitz: Bureau of Reclamation

Academia

- Yoram Cohen, PhD: Professor at UCLA, Director of the Water Technology Research Center
- John Lienhard, PhD: Professor at MIT, Director of World Water and Food Security Lab and Center for Clean Water and Clean Energy

10 BERKELEY LAB | U.S. DEPARTMENT OF ENERGY Energy Efficiency & Renewable Energy ADVANCED MANUFACTURING OFFICE | ENERGETICS A Subsidiary of VSE Corporation

Scope of Desalination System Energy-Water Bandwidth Study

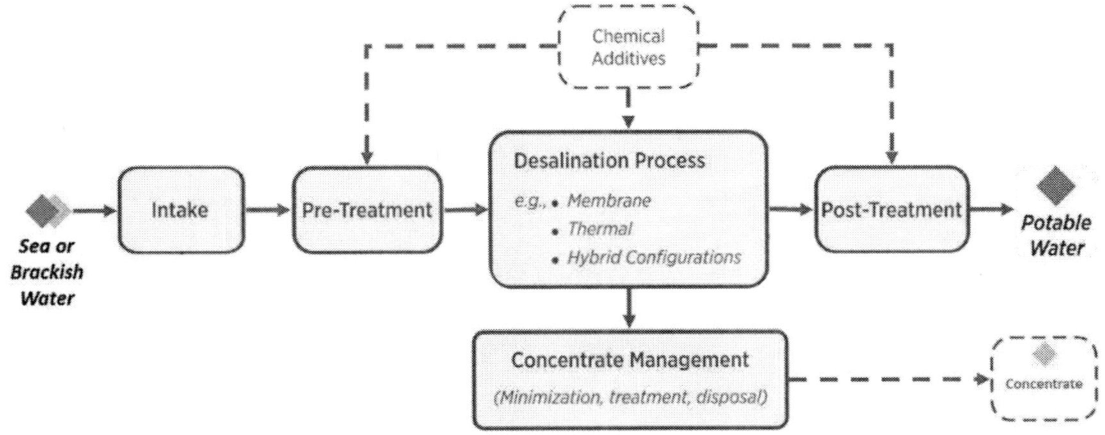

- Intake conditions: Saline water (focus is seawater)
- Product water: potable water
- System encompasses intake through product water leaving plant boundary

BERKELEY LAB | U.S. DEPARTMENT OF ENERGY Energy Efficiency & Renewable Energy ADVANCED MANUFACTURING OFFICE | ENERGETICS A Subsidiary of VSE Corporation

Boundary Analysis Framework – Whole System

Whole System Economic and Operational Characteristics

Fixed Cost ($)	Construction	Siting	Permitting	Land Use	Equipment	Construction time	
Variable Cost ($)	Labor	Chemicals	Maintenance	Parts	Disposal		
Energy Cost ($)	Electricity	Natural Gas	Steam				
Operational characteristics	Water quality	Flow rates (m^3/day)	Temperature (°C)	Pressure (PSI)	Energy (kWh, MMBtu)	Chemicals	Brine (m^3/day)
	System Lifetime	Modulation capability	Operation when under loaded	Capacity Factor	Availability (hours/year)	Operability with other components	Water recovery (%)

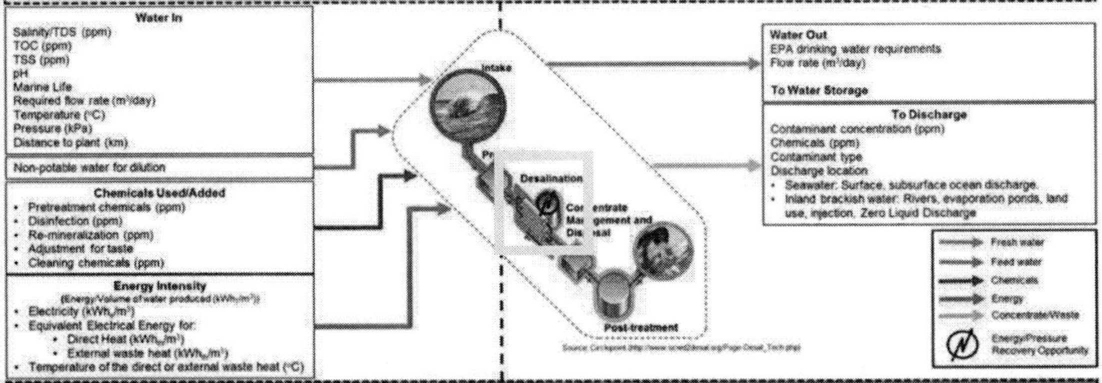

Whole System Environmental Considerations

Emission Intensity (tCO_2/G)	• CO_2 emissions (tCO_2/m^3 of freshwater produced)
Other	• Marine life • Concentrate disposal • Consumables disposal and chemical use/disposal • Electric grid impact • Water system integration • Co-location with power generators

BERKELEY LAB | U.S. DEPARTMENT OF ENERGY Energy Efficiency & Renewable Energy ADVANCED MANUFACTURING OFFICE | ENERGETICS A Subsidiary of VSE Corporation

Boundary Analysis Framework – Unit Operation

Whole System Economic and Operational Characteristics								
Fixed Cost	Construction	Equipment	Permitting					
Variable Cost	Chemicals	Maintenance	Membrane or vessel replacement		Labor			
Energy Cost	Energy for water recovery		Pressurization	Conveyance				
Operational characteristics	Feedwater type	Flow rate (m^3/day)	Energy source	Typical size (m^2)	Pressure (PSI) or TDH (ft)	Temperature (°C)	Quantity and type of energy available for recovery	
	Energy consumption		Frequency of backwash and system cleaning		Typical water recovery (% of feedwater)		Dispatch-ability	System lifetime

Water from Pre-treatment
Salinity/TDS (ppm)
TOC (ppm)
TSS (ppm)
pH
Flow (m^3/day)
Temperature (°C)
Pressure (kPa)

Chemicals Used/Added
- Cleaning chemicals (ppm)
- Process specific solutions (e.g. Draw solution NH_4HCO_3)

Energy Intensity
(Energy/Volume of water produced (kWh_T/m^3))
- Electricity (kWh_e/m^3) for:
  - Conveyance or,
  - Pressurization
- Equivalent Electrical Energy for:
- Direct Heat (kWh_{th}/m3)
- External waste heat (kWh_{th}/m3)
- Temperature of the direct or external waste heat (°C)

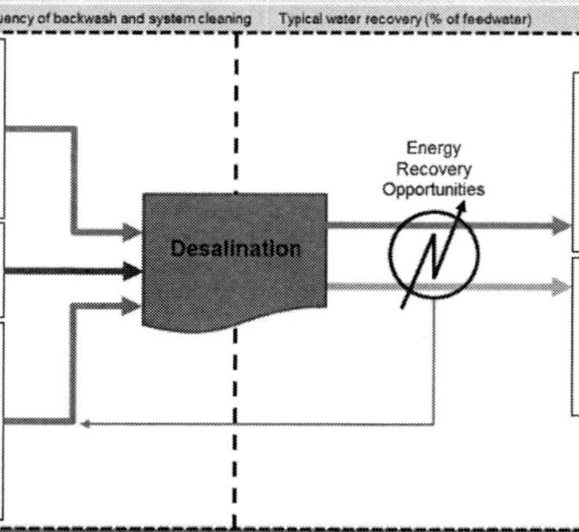

To Post-treatment Stage
Salinity/TDS (ppm)
pH and Alkalinity
TOC (ppm)
Flow rate (m^3/day)
Temperature (°C)
Pressure (kPa)
Dissolved gases or chemicals (e.g. Ozone ppm)

To Disposal Stage
Pressure (PSI)
Temperature (°C) in case of waste heat recovery
Enthalpy (Energy/Volume)
Contaminant concentration (ppm)
Contaminant type
Flow rate (m^3/day)

Environmental Considerations	
Emission Intensity (mtCO2/m³)	• Source of electricity generation • Source of thermal generation, including waste heat
Other	• Electric demand • Time of use considerations

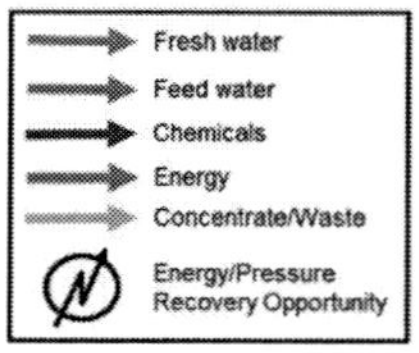

Desalination Technologies Reviewed in Volume 1

Seawater

Intake	Pre-treatment	Desalination	Post-treatment	Concentrate Management
• Open-ocean intake - Screened • Sub-surface intake - Beach wells - Offshore radial collector wells	• Membrane filtration - Microfiltration - Ultrafiltration • Media filtration • Sand filtration • Cartridge filtration • Disc filtration • Flocculation • Sedimentation • Chlorination • Dissolved air flotation	• Reverse osmosis • Forward osmosis in combination with RO or a thermal technology • Multi-stage flash distillation • Multi-effect distillation • Mechanical vapor compression • Thermal vapor compression	• Remineralization • Disinfection • Boron removal	• Surface water discharge • Zero liquid discharge - Brine concentration - Crystallization

Brackish Water

Intake	Pre-treatment	Desalination	Post-treatment	Concentrate Management
• Sub-surface intake	• Media filtration • Cartridge filtration	• Reverse osmosis • Electrodialysis and Electrodialysis Reversal • Nanofiltration	• Remineralization • Disinfection • Chlorination • Boron removal	• Surface water discharge • Evaporation ponds • Deep well injection • Zero liquid discharge

Total 2015 Worldwide Installed Desalination Capacity By Feedwater Type

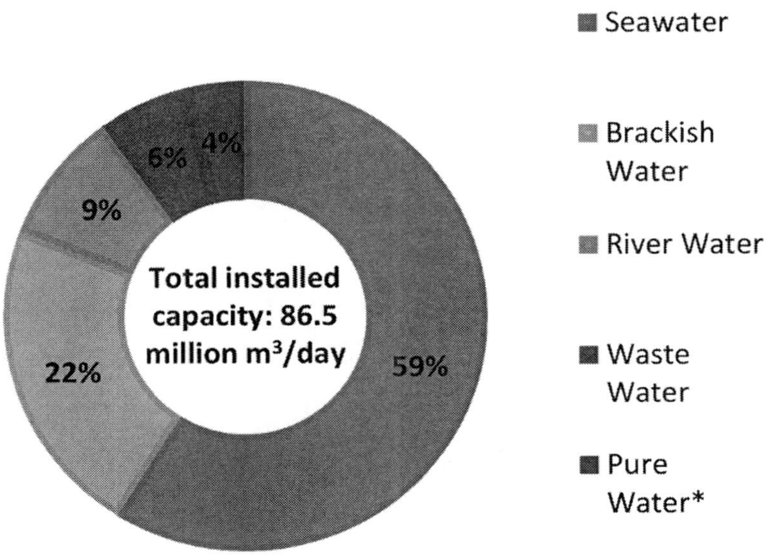

* Pure water sources are desalinated to achieve ultrapure water recovered for certain processes

15

Data source: Virgili, F., Pankratz, T., and Gasson, J. (2016). IDA Desalination Yearbook 2015–2016

Total 2015 Worldwide Installed Capacity by Technology

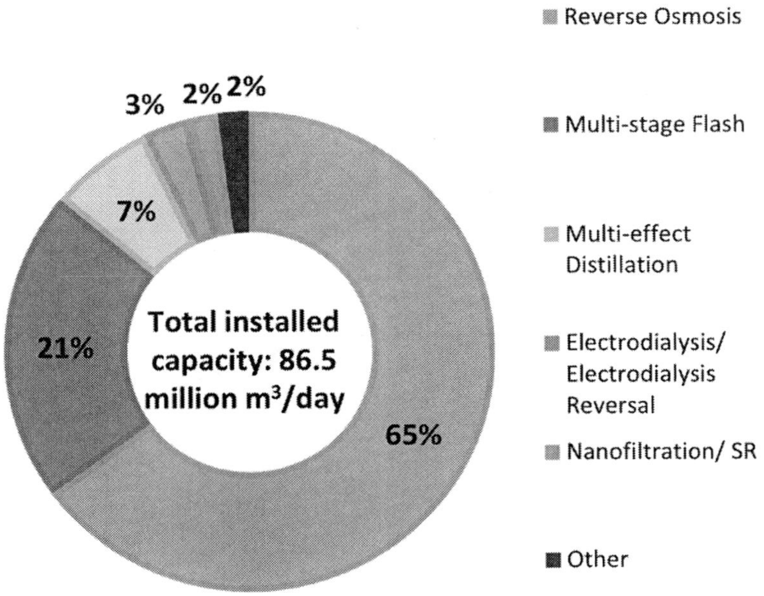

16

Data source: Virgili, F., Pankratz, T., and Gasson, J. (2016). IDA Desalination Yearbook 2015–2016

Preliminary Energy Data (Seawater)

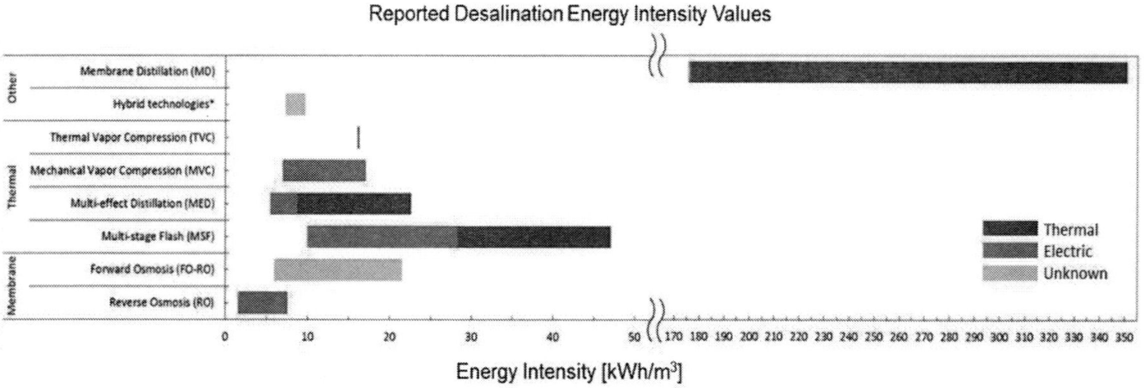

Key Finding: Reported energy data needs further refinement before being used

- Values need to be reported with operational characteristics, e.g. intake and product water flow rate, recovery and plant size, salinity, and temps (for thermal processes), desalination unit operation, use of energy recovery or waste heat
- Addition of electrical and thermal must account for generation losses associated with converting thermal energy to work

17

BERKELEY LAB | U.S. DEPARTMENT OF ENERGY | Energy Efficiency & Renewable Energy | ADVANCED MANUFACTURING OFFICE | ENERGETICS A Subsidiary of VSE Corporation

Energy Accounting, Carbon Dioxide Emissions

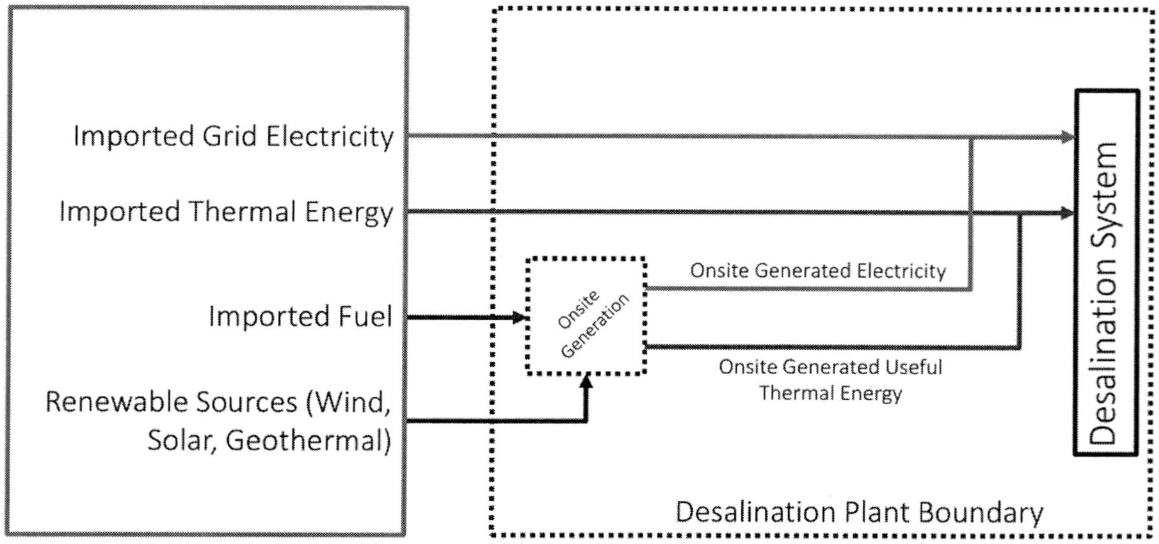

Carbon Emissions Depend on the Energy Source

18

BERKELEY LAB | U.S. DEPARTMENT OF ENERGY | Energy Efficiency & Renewable Energy | ADVANCED MANUFACTURING OFFICE | ENERGETICS A Subsidiary of VSE Corporation

Example SWRO Desalination Costs

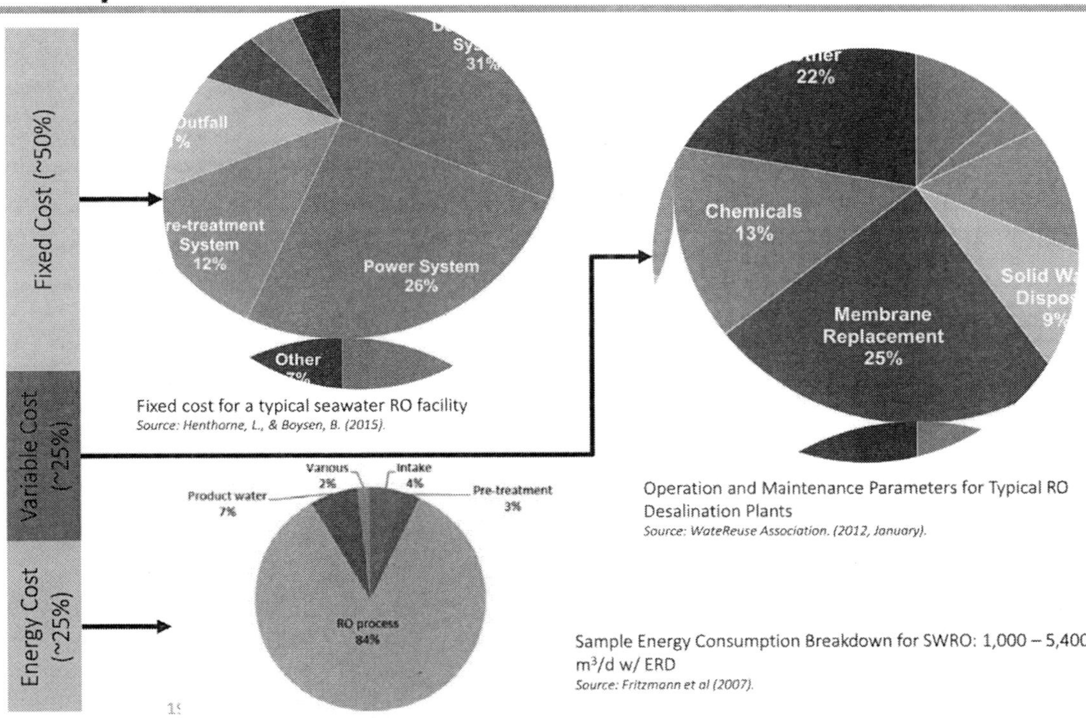

Fixed cost for a typical seawater RO facility
Source: Henthorne, L., & Boysen, B. (2015).

Operation and Maintenance Parameters for Typical RO Desalination Plants
Source: WateReuse Association. (2012, January).

Sample Energy Consumption Breakdown for SWRO: 1,000 – 5,400 m^3/d w/ ERD
Source: Fritzmann et al (2007).

Scenarios

Seawater Scenarios

- Supplying all populations within a certain distance (e.g., 10 miles) of a seawater coastline
- Supplying (all or a percentage of) public water for all water stressed regions (current and projected out to a given year) in the U.S. through seawater desalination
- Supplying all populations with existing water costs that are above the existing and estimated potential water cost produced through seawater desalination
- Supplying regions with large renewable energy resources, but insufficient energy storage capability
- Supplying all populations within the water distribution network of retired coal or operating thermal power plants
- Supplying all U.S. public water supply through seawater desalination (offers theoretical maximum for a given energy intensity)

Brackish Water Scenarios

- Supplying all population centers of over one million people with nearby saline water resources and concentrate disposal options using existing practices
- Supplying (all or a percentage of) public water for all water stressed regions (current and projected) in the U.S. where brackish water sources exist
- Supplying all populations with existing water costs that are above the existing cost of water produced through brackish water desalination.

Factors for Developing Scenarios

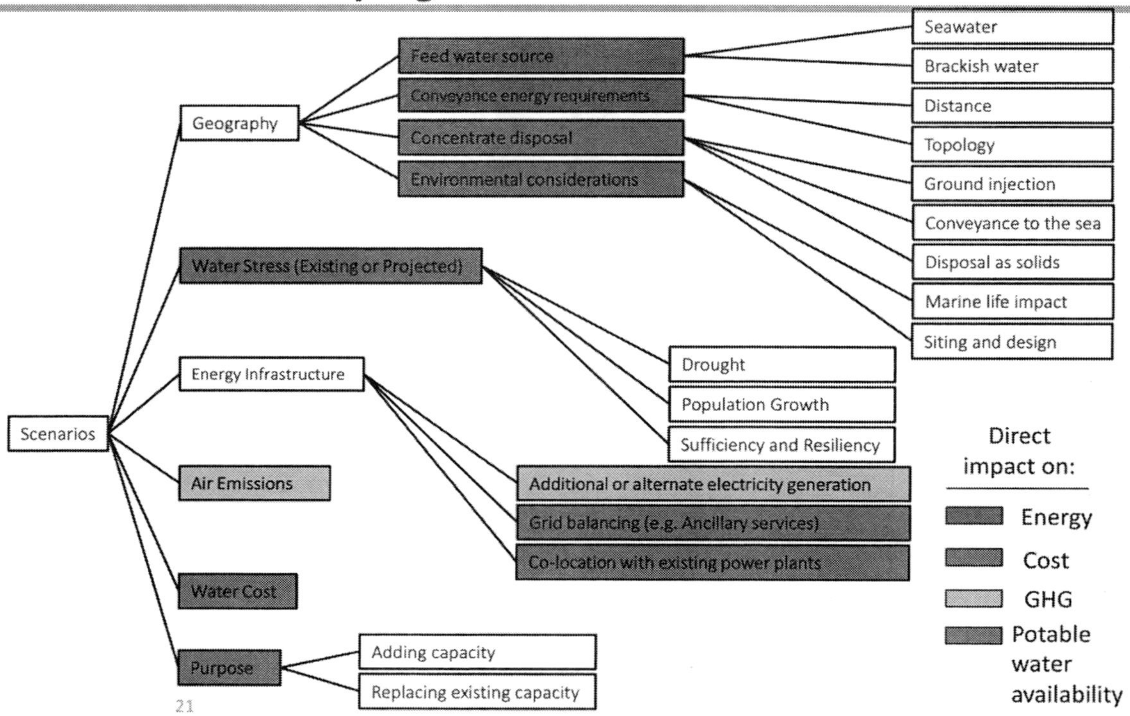

21

Key Takeaways Based on Volume 1

- **Energy intensity data reported in literature is seldom reported in a consistent manner (e.g. consistent system boundaries and input/output water quality) and must be harmonized**
- **Determining the best technology options and the energy intensity for a given desalination system is complex and will depend on many factors including:**
  - **Feedwater characteristics, output water quality, desired recovery, concentrate disposal options**
- **For desalination technologies, aside from RO:**
  - **Forward Osmosis in combination with another desalination process is featured prominently in the literature for seawater desalination.**
  - **Electrodialysis is already established and competitive against RO for brackish water desalination.**

22

BERKELEY LAB | U.S. DEPARTMENT OF ENERGY | Energy Efficiency & Renewable Energy | ADVANCED MANUFACTURING OFFICE | ENERGETICS A Subsidiary of VSE Corporation

Key Takeaways Based on Volume 1 (cont.)

- **Energy intensity for RO membranes has been reduced over the years and is approaching thermodynamic limits for crossflow RO desalination**
  - **Advances in pre-treatment, high pressure pumping, pressure energy recovery, improved operational practices, and development of fouling resistant membranes may offer further opportunity for energy reduction for the desalination system**

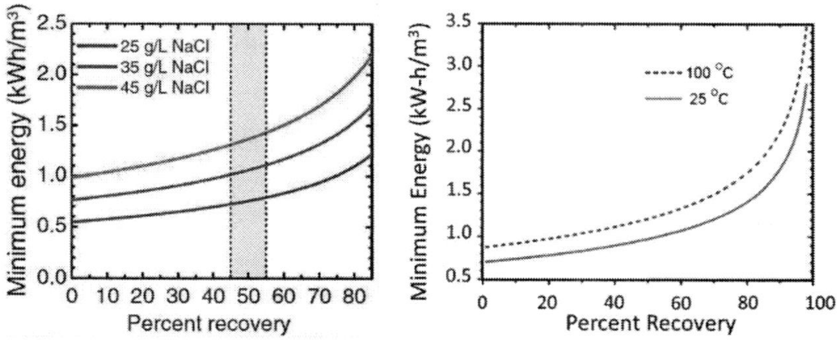

Figures: Elimelech and Philip. 2011; National Research Council. (2008)

Conclusions

- **An *Energy Bandwidth Study* is being performed to analyze energy use and saving opportunities (and carbon emissions and savings opportunities) for desalination systems in the U.S. to produce potable water, with a focus on seawater desalination**
- **Energy data are broken down by unit operation to identify processes and energy-saving technologies with the greatest opportunities for energy savings**
- **Volume 1 of the study was undertaken to provide background research and a framework for Volume 2, which is underway. Volume 2 is expected to be available in 2017 and will provide the bandwidth results of energy and CO_2 consumption and savings opportunities.**

> **Results presented in this talk are <u>draft data</u>; Vol. 1 & 2 studies are peer reviewed. If you are interested in receiving updates, please contact me!**

Thank you!

Lawrence Berkeley National Laboratory

Prakash Rao – Technical Lead

(prao@lbl.gov)

Energetics Incorporated

Caroline Dollinger

(cdollinger@energetics.com)

AMO

Joe Cresko - Strategic Analysis Technology Mgr.

(joe.cresko@ee.doe.gov)

25

Identified Technology Questions

- **Maturity level and performance of energy recovery devices, including maintenance and component lifetime**
- **Options for reducing mineral scaling, fouling, and concentrate polarization for ultrahigh permeability membranes**
  - **Target flux/surface area of membrane to achieve water recovery rates in line with meeting a total cost of water equivalent to the cost from ground/surface freshwater sources**
- **Challenges to developing fouling, oxidation, and/or scale resistant membranes**
- **Opportunities for greater performance of pressurization pump and motor, including advanced motor technologies to meet pump requirements**

26

Identified System Integration Questions

- **Barriers to desalination as a means for greater recovery and direct (or indirect) reuse of municipal wastewater**
- **Barriers to integrating renewable energy sources with desalination facilities**
- **Can desalination as serve as a "water battery" and enable: 1) better integration of renewable energy sources for grid balancing and 2) lowering the total cost of water by creating additional revenue mechanisms through demand response**

Identified Cost Questions

- **Methods for reducing manufacturing costs of desalination system components, including: first time cost for RO membranes, extending membrane lifetime and maintaining design throughput, reuse of cartridge filters**
- **Reducing capital costs of staged RO desalination**
- **Possibility of financing mechanisms for lowering desalination amortization costs analogous to the financing schemes that have in-part enabled greater uptake of solar energy**
- **Effect of increasing water source diversity, and subsequently improving water infrastructure resiliency, when comparing total water costs across water sources (i.e., what is the cost if Los Angeles lost partial access to water for one summer?)**
- **Comparison of environmental impacts of intake and concentrate disposal to fresh water use**

Key Takeaways Based on Volume 1 (cont.)

- **Chemicals used in pre-treatment can reduce the energy intensity of the desalination process, but need to be recovered and/or safely disposed**
- **Integration of renewable energy and water storage, particularly to build desalination systems to support and balance the electric grid and reduce CO2 emissions, is promising**

29

459842 Micro-Deformation in Polyethersulfone Inner-Selective Thin Film Composite Hollow Fiber Membranes Under High Hydraulic Pressures in Pressure Retarded Osmosis Process

Abstract:

Micro-deformation in Polyethersulfone Inner-selective Thin Film Composite Hollow Fiber Membranes under High Hydraulic Pressures in Pressure Retarded Osmosis Process

Wenxiao Gai, Xue Li, Jun Ying Xiong, Chun Feng Wan, Tai-Shung Chung *

Department of Chemical and Biomolecular Engineering, National University of Singapore, 4 Engineering Drive 4, Singapore 117585, Singapore

* Corresponding author: Tel.: +65 6516 6645. E-mail: chencts@nus.edu.sg

Abstract

Pressure retarded osmosis (PRO) is one of the promising methods to harvest the osmotic energy , which is a clean and renewable candidate to mitigate the dependence of mankind on non-renewable and environmentally detrimental fossil fuels, by separating the two solutions using a semi-permeable membrane. Recently, lots of thin film composite (TFC) hollow fiber (HF) membranes have been developed for PRO application because of their self-support configuration and high packing density properties. This work may be one of the first attempts to systematically investigate (1) the micro-deformation of the inner-selective TFC HF membranes under high hydraulic pressures for osmotic power generation and (2) the evolution of permeation properties with the increased hydraulic pressures applied in the lumen side from 0 to 20 bar. Experimental results show that pre-stabilization at 20 bar close to the burst pressure of the polyethersulfone (PES) TFC HF membranes could improve the water flux and power density dramatically due to the increased membrane surface area, stretched polyamide selective layer and decreased membrane structure parameter. It was also proved by the intermittent cycle tests that the enhancement of water flux and power density is sustainable after pre-stabilization at 20 bar without compromising the selectivity. These results may provide significant technical implications to design desirable membranes for PRO application and maximize the output of the PRO process.

453222 Anitifouling Polyimide Membrane with Surface-Bound Silver Particles

Abstract:

To improve antifouling and antibacterial properties of the polyimide (PI) membrane, silver particles are grafted on the membrane surface via a three-step modification route. The reaction mechanism is confirmed by Fourier Transform infra-red analysis (FTIR) and physical characterizations of the modified PI membranes. The anti-biofouling performance is evaluated by disk diffusion test and bacterial suspension immersion experiment. The antifouling effect of the silver-modified membranes is also assessed in the bovine serum albumin (BSA) aqueous solution. Both the anti-biofouling and antifouling properties of the membrane are greatly improved by the silver modification. Effects of the thermal treatment temperature on the property and silver stability of modified PI membrane are also investigated systematically.

466933 Green Modification of Outer Selective P84 Nanofiltration (NF) Hollow Fiber Membranes for Cadmium Removal

Abstract:

Green Modification of Outer Selective P84 Nanofiltration (NF) Hollow Fiber Membranes for Cadmium Removal

Jie Gao, Shi-Peng Sun, Wen-Ping Zhu, Tai-Shung Chung*

Department of Chemical and Biomolecular Engineering

National University of Singapore, Singapore 117576

*Corresponding author

Tel: +65-65166645; fax: +65-67791936; Email: chencts@nus.edu.sg

Abstract:

Nanofiltration (NF) is a membrane filtration process with a membrane pore size in the sub-nano scale. Due

to its special rejection mechanisms: size exclusion and charge repulsion, NF is able to offer good rejections to multi-valent heavy metal ions without much compromising water fluxes as compared to those of reverse osmosis (RO) membranes.

There is a growing interest in thin-film composite (TFC) NF hollow fiber membranes. The TFC layer can be formed on both inner and outer surfaces of hollow fiber membranes. The outer selective TFC hollow fiber membranes may have less mass transfer resistance and lower chances being blocked by foulants compared to the inner ones. Thus, the outer selective TFC fibers can be fabricated into smaller fiber dimensions with a higher package density in modules. Despite these advantages, the molecular design of an ideal TFC layer on the outer surface of hollow fiber membranes remains a challenge, making the mass production difficult. Since most TFC layers are made from interfacial polymerization of m-phenylenediamine (MPD) and trimesoyl chloride (TMC), removing excess solution after the MPD coating is the major obstacle in the process. Vacuum assist interfacial polymerization or wipe drying are required to make a defect free polyamide layer. Another obstacle is that a large amount of alkane solvents has to be used during interfacial polymerization, which brings extra costs and environmental issues. Eliminating alkane solvents is critical to lower manufacture costs and yield a green production process.

In this study, green methods to prepare the outer selective TFC hollow fiber membranes were explored by firstly modifying the membrane substrate with polyethyleneimine (PEI) and then by water soluble small molecules such as glutaraldehyde (GA) and epichlorohydrin (ECH). Using P84 polyimide as the substrate, not only do these modifications decrease substrate's pore size, but also vary surface charge by making the membranes less positively charged. As a result, the resultant membranes have higher rejections against salts such as Na_2SO_4, NaCl and $MgSO_4$. The PEI and then GA modified membrane has the best separation performance with a NaCl rejection over 90 % and a pure water permeability of 1.74 ± 0.01 $Lm^{-2}bar^{-1}h^{-1}$. It also shows an impressive rejection of 94 % for $CdCl_2$ removal during long-term stability tests. The $CdCl_2$ rejection remains higher than 90% when testing it from 5 to 60 °C. This study may provide useful insights for green manufacturing of next-generation NF hollow fiber membranes.

Acknowledgements:

The authors would like to thank Chinese Gansu province for the research project entitled "Membrane development to removal heavy metals" with the grant number of R-279-000-360-597 and GSK-EDB Trust Fund for the project entitled "New membrane development to facilitate solvent recovery and pharmaceutical separation in pharmaceutical syntheses" with the grant number of R-706-000-019-592. The authors also acknowledge Mr. Z.L. Cheng, Dr. X. Li, Ms, Y. Zhang, Dr. Z.Z. Zhou and Dr. Y.K. Ong for all their helps and suggestions on the experimental work. Thanks are also due to Mr. K.P. Ng for his help on the NF set-up.

Chevron

human energy

Sustainable Development and Footprint Optimization in Shale Development

Janet Peargin, Footprint Program Manager
Hong Jin, Lead HES Engineer

Chevron Energy Technology Company

November 17, 2016

The challenge
footprint of shale/tight rock development

Many wells & moving parts

Footprint changes over time

- Jobs
- Community GDP
- Desired energy

- Land/habitat
- Emissions
- Trucks
- Noise
- Waste
- Resource competition

Chevron's environmental principles

- Include the environment in decision making
- Reduce our environmental footprint
- Operate responsibly
- Steward our sites

Footprint optimization: everybody plays a role

We can't solve problems by using the same kind of thinking we used when we created them.

– Albert Einstein

Shifting your perspective is so much more powerful than being smart.

– Astro Teller, Google "X" 2016 TED Talk

The opportunity: sustainable development

SEE across the asset life to create value and protect people and the environment

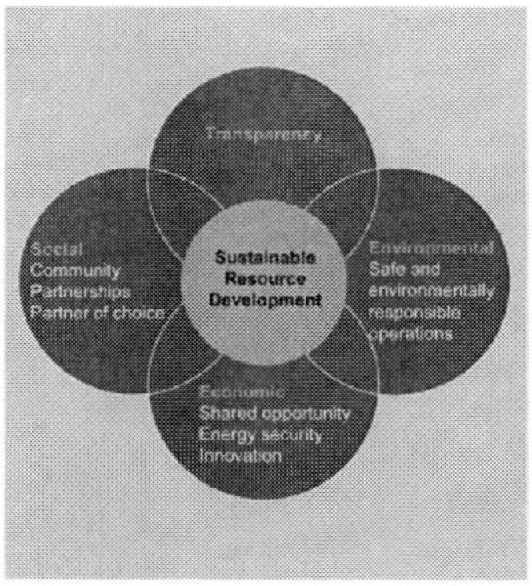

SEE triple bottom line value

SEEk win-win (all-win) ideas
Social, Environmental, Economic

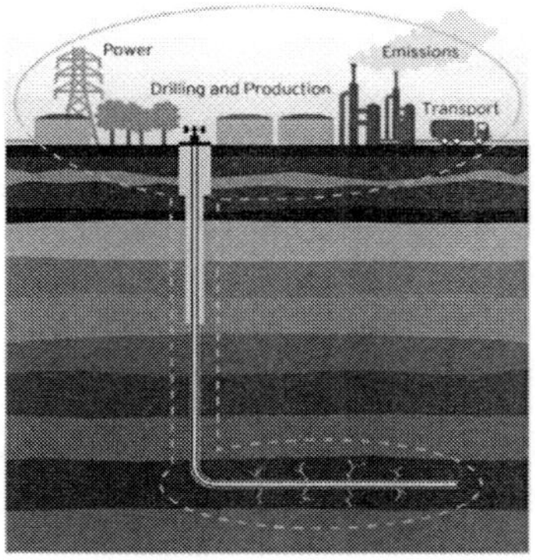

Optimize footprint

Integrated system of materials and energy that changes over time

Design with the end in mind

Decommission, restore, reuse

The win-win (all-win): optimize footprint to enhance value and reduce cost/risk

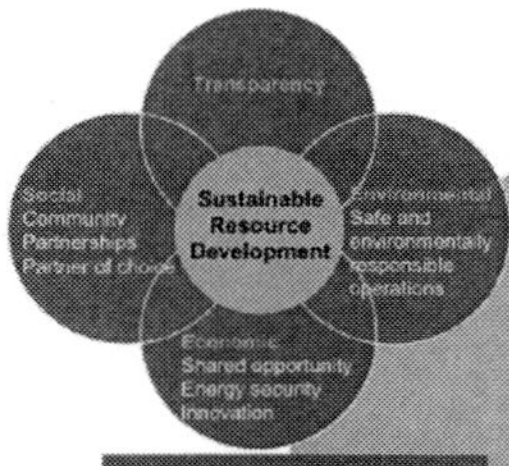

Transparency

- Compact AND safe
- Conserve water and natural resources
- Preserve habitat and biodiversity
- Reduce waste, energy, noise, air emissions
- Optimize logistics and truck traffic
- Provide net benefit to community

Social | **Environmental**

achieved through

1. **Integrated, long-term plans.** Frame for SEE drivers
2. **Informed decisions.** Include SEE drivers. Use ISO LCA*
3. **Good design.** Mobile, compact, flexible, efficient

*Life Cycle Assessment Standard 14040-14044

Economic

Improve planning: SEE through the lens of sustainable development and SEE win-win (all-win) opportunities

Look at the big picture (asset, region)
Look at the long-term and change over time
Include diverse stakeholder perspectives

Take a systems view. Integrate across functions, activities, hydrocarbon/water/power
Plan and design for social, environmental, and economic (SEE) value drivers

Consider:

- Pace and scale of development
- Sequencing of events
- Degree of centralization
- Logistics and supply chain
- Ability to share infrastructure
- Power/fuel options
- Integration points

Improve decision quality: SEE more broadly and longer-term

Compliance is expected.

A traditional financial analysis is narrow and short-term.

SEE more broadly and longer term to identify risks and opportunities … and enhance decision quality.

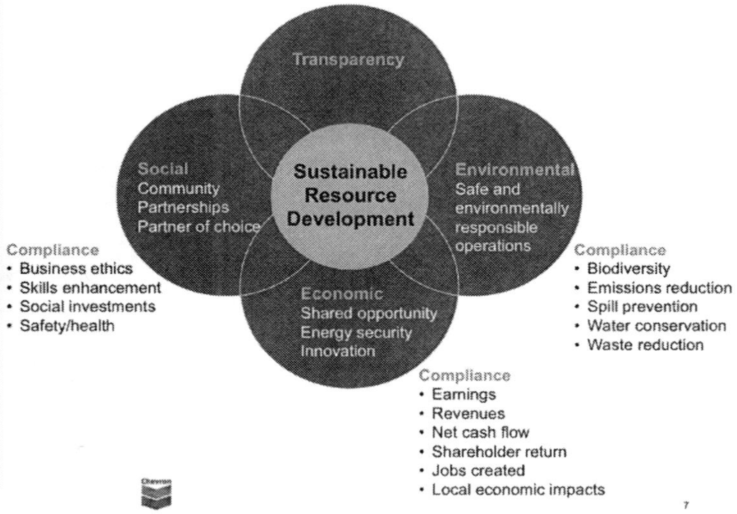

Improve decision quality: quantify environmental tradeoffs with ISO life cycle assessment (LCA)

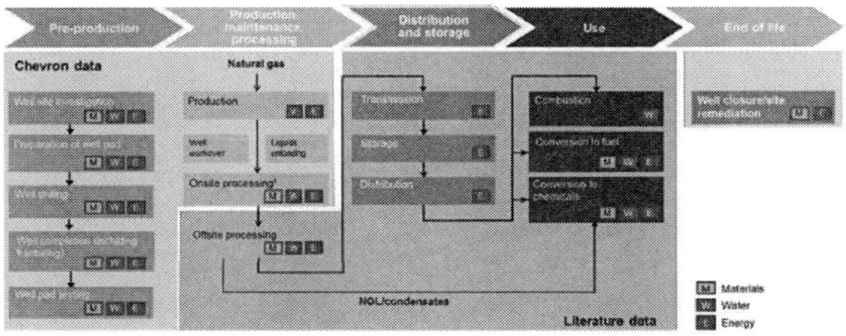

- What is our environmental footprint?
- Where are our big sources?
- What is the "best" fuel for power?
- What if I use this technology vs. that?
- What are the on and offsite impacts?

Improve design: apply win-win (all-win) principles

Plan/design for sustainable development	• Develop long-term integrated plans • Frame for SEE (social, environmental, economic)
Integrate and innovate	• Look across the asset life at hydrocarbons/water/power & different activities (drilling, operations, etc.)
Question business as usual	• What do stakeholders want? Is there a better way?
Do *more* with *less*	• Reduce – reuse – recycle – be (energy/materials) efficient • Eliminate no-value steps & equipment
Disturb less land	• Fewer pads; share infrastructure/right-of-ways; preserve habitat corridors
Conserve water	• Use less & lower quality; treat to quality needed for use; consider reuse
Reduce waste	• Flares/emissions, discharges, energy use, no-value steps/equipment
Green the process	• Consider non-toxic chemistry, power/fuel efficiencies, fewer steps
Leverage technology	• Look to downstream, offshore, suppliers, competitors, others • Think mobile, modular, compact, flexible, prefab, energy efficient, automated, standard
Leverage supply chain	• How can suppliers help? Can we hire and source locally?
Make informed decisions	• SEE the big picture. Consider short- *and* long-term costs and value. • Use life cycle assessment to quantify environmental aspects of alternatives

Win-win example: electricity from field gas

Better emissions/safety/reliability/production/cash flow

Problem

Need cost-effective, low-emissions power source at remote site

Solution

Micro-turbines make electricity from field gas*

Win-Win Do *more* with *less*

MORE

Safety, flexibility, schedule certainty, reliability (8–10%), production/cash flow (sooner)

LESS

Equipment, permitting, O&M, trucks, emissions, material transfer/spill potential

* *Great for ESPs. Rod pumps doable with more systems engineering*

Everyone plays a role: ask questions

- What are our stakeholder concerns?
- What perspectives are we missing?
- What are our footprint boundaries?
- What are the priorities and tradeoffs?
- What does good (enough) look like?
- How can we minimize environmental and community impacts?
- How do we do more (product) with less (material, equipment, waste, etc.)?
- How do we design for change over time to operations, region, community?
- What can we learn from other processes?
  - LeanSigma®
  - Environmental, social, health impact assessment
  - Energy Efficiency and Waste Minimization assessments

LeanSigma is a federally registered service mark of TBM Consulting, Inc.

11

Key takeaways

- We all play a role in creating our company footprint – a system of materials and energy that changes over time
- Think beyond compliance. Include SEE (social, environmental, economic) drivers in planning, designs, and decisions.
- SEE through the lens of sustainable development to SEE opportunities. SEE the difference.
  - Cost savings from less equipment, optimized layout/network
  - Win-win ideas to save $ AND protect people & the environment
  - Decisions that are more robust
- A financial analysis is only part of the risk/uncertainty. We <u>can</u> quantify and include SEE drivers in decision-analysis.
- Sustainable development is good business. SEE it reduce cost and risk, add value, and improve performance.
- Everyone plays a role. You can contribute. Just change your perspective to SEE the win-win (all-win) opportunities to do more (product, energy) with less (energy, materials, waste, emissions, trucks, etc.)

12

References

- Hardisty, P.E. *Environmental and Economic Sustainability.* CRC Press, New York. ISBN 978-14200594891-56679-643-2, 378 pp. 2010.

- Sivapalan, M.; WorleyParsons. *Sustainable Development of Unconventional Oil & Gas Resources.* Prepared for presentation at *GasTech* 2012 Conference, London, 2012.

- Tanzil, D.; Dragan, M.; Bourque, B.; Seel, K.; Taylor, T.; Golder Associates. *Evaluating Social and Environmental Risk and Opportunities in Site Selection and Routing.* Copyright 2015, Society of Petroleum Engineers. SPE-173488-MS. Prepared for presentation at the SPE E&P Health, Safety, Security, & Environmental Conference – Americas held in Denver, Colorado, USA, 16–18 March 2015.

- Thomas, Charlotte. *Industrial Ecology: Moving Beyond Traditional Engineering Borders.* Society of Women Engineers magazine, spring edition, pp. 24-30. 2010

13

461852 Green Net Value Added As a Sustainability Metric Based on Life Cycle Assessment: An Application to Bounty ® Paper Towel

Abstract:

Sustainability measurement in economics involves evaluation of environmental and economic impact in an integrated manner. In this study, we combine system level economic data with environmental impact from a life cycle assessment (LCA) of a common product. We are exploring a costing approach that captures traditional costs and incorporates externality costs to provide a convenient, easily interpretable sustainability metric. Green Net Value Added (GNVA) is a type of full cost accounting that incorporates total revenue, the cost of materials and services, depreciation, and environmental externalities. It adjusts the traditional economic value added for environmental externality costs. Two, but not all, of the potential environmental impacts calculated by the standard LCIA method (TRACI) could be converted to externality cost values. We compute externality costs disaggregated by upstream sectors in order to accurately reflect the externality costs associated with different industries along the supply chain. We preset full cost, green profit and GNVA to evaluate the relative sustainability of Bounty® paper towels manufactured at two production facilities. We found that the longer running, more established line had a higher GNVA than the newer line. The dominant factors contributing to externality costs are to come from the stationary sources in the supply chain: electricity generation (27-35%), refineries (20-21%), pulp, and paper making (15-23%). Health related externalities from Particulate Matter (PM2.5) and Carbon Dioxide equivalent (CO_2e) emissions appear largely driven by electricity usage and emissions by the facilities, followed by pulp processing and transport. Supply chains for other facility purchases, product distribution, and end-of-life management (landfilling/incineration) contribute the least to the environmental health-related costs.

474653 Sustainability Metrics for Solar Powered Charging Infrastructure for Electric Vehicles

Abstract:

This presentation will include an analysis of sustainability metrics associated with the installation of solar panels in parking lots designed to provide electricity for commercial use and shade for those who park under the solar panels. The installation will also include electric vehicle supply equipment to charge batteries in electric vehicles. The system is connected to the electrical grid so that the electricity generated can be used efficiently and electric vehicle batteries can be charged at all times. Important sustainability metrics are cost of electricity generated, cost of vehicle charging, value of shaded parking, convenience of vehicle charging, land area, water, air quality, greenhouse gas emissions, and materials for the installation and equipment. Social metrics include impact on health status of the community, psychological well being, harmony, values, agency, and environmental justice.

458893 Evaluation of Low-Energy Recovery of N-Methyl-2-Pyrrolidone Solvent from Waste Streams in Specialty Resin Manufacture

Abstract:

N-methyl-2-pyrrolidone (NMP) is a common solvent used as a reaction medium in the manufacture of specialty polymeric resins and their precursors. NMP is used primarily because of its thermal stability, dipolar aprotic properties, and ability to produce a high molecular weight resin precursor. Since NMP is not consumed by the reaction, an aqueous hazardous waste stream is generated. A previous case study at a DuPont specialty chemical plant by the authors, shows that a distillation recovery system that can effectively purify NMP from the waste stream. The distillation recovery process reduces operating cost by 83%, but only reduces the life cycle carbon footprint by 40%. This is due to the high energy demand of the distillation column. Additionally, high NMP purity levels in the resin manufacture process and the complex multicomponent nature of the waste stream, required a more extensive design for the recovery operation.

Alternative distillation designs were explored, but none reduced energy consumption and life cycle carbon emissions as significantly as the economic life cycle impact. More sustainable low energy separations were investigated. Various membrane, adsorption, and extraction separation strategies have the potential to reduce energy demand, but each has its own limitations. An extraction operation was selected as the best candidate for separation, and the proposed process considers alternative uses for the recovered NMP waste in other manufacturing scenarios.

Results with a chloroform extraction operation to produce NMP of suitable quality for alternative use, provides the best option for solvent recovery. The new process can not only yield environmental life cycle improvements, based on various impact factors, but provide economic benefits as well. Yearly projections of waste reductions, life cycle improvements, energy utilization, and cost savings are presented. Life cycle carbon emissions are reduced by over 60% compared to the base case. Since NMP is a prominent solvent for several commercial sectors, methods aimed at reducing waste help promote broader sustainability goals. This study shows how sustainability metrics can be used to guide decision making when evaluating other processing alternatives and uses of the final product. (This work is partially supported by a grant from the EPA Pollution Prevention program).

468272 Thermoelectric Power Technology Choices Based on Water Availability

Abstract:

Thermoelectric Power Technology Choices Based on Water Availability National Energy Technology Laboratory (NETL) has developed a prototype water-energy model to investigate issues surrounding the energy-water nexus as it relates to the effect of power generation plant design choices on water availability and the overall costs of supplying water for thermoelectric power generation.Â The model utilizes the database developed by a team of researchers with the Water Security Program at Sandia National Laboratory (SNL). The database includes water availability, cost, and water use for electric power, agriculture, municipal and industrial for 38 conterminous states representing various Interconnects within the U.S. Data on current and future water use for thermoelectric power are derived from NETL databases on water use by technology and projections from the Energy Information Administrationâ€™s National Energy Modeling System (NEMS). Current and forecasted water availability data from the World Resources Institute are used to develop three water availability scenarios. The prototype water-energy model is used to project the amount of water available for power generation, to project water demand for future power generation based on a NEMS forecast, to identify regions that experience potential water shortfalls, and to estimate the cost of meeting water demand where water shortfalls exist for each water availability scenario. A case study of the Brazos River Basin in Texas is used to illustrate a specific application and to highlight the capabilities of the prototype water-energy model.

464240 Deciphering Lignin Deconstruction Mechanism of Lower Termite

Abstract:

Although fascinating, the exact process of plant cell wall degradation developed by nature and implemented in lower termites remains unclear. We have been conducting systematic studies on the lignin deconstruction mechanism over the last seven years as an effort to develop nature-inspired technologies for lignocellulose processing. A series of experimental and analytical techniques from various disciplines were used in our studies. Scientific questions explored included mircrochemical environment of the termite guts, structural and composition changes of lignocellulose along the termite digestion track, critical enzyme systems involved in lignin deconstruction, chemical reactions that modifies lignin structure, and disassociation energy levels of different chemical bonds that form lignin structure. Integrating these results makes it possible to propose a chemical reaction mechanism and associated lignin degradation pathways. The mechanism and the pathway will be presented and discussed in the proposed paper. The results of the study provide new insights for developing more sustainable processes for plant cell wall deconstruction required in biorefineries.

455682 Production of Food, Energy, and Fuel from Microalgae in Holograhic Diffractive Optic-Solar Glass Reactors

Abstract:

This project is focused on developing technology platforms for the solar-assisted production of food, fuel, and clean water from municipal wastewater sources. This novel food, energy water (FEW) platform employs new light management strategies to simultaneously: (1) generate electrical energy; and (2) optimize algae growth for food and fuel. This is accomplished by use of a holographic diffractive optic-solar glass (HDO-SG) reactor. The glass transmits the visible light spectrum that is used for photosynthesis and internal reflects the IR and UV light that is directed toward photovoltaic (PV) cells that are integrated into the reactor. The PV- generated energy can be used to operate the photobioreactor. Preliminary results show that *Chlorella* sp and *Scendesmus* sp. algae grow well in reactors containing HDO-SG. Water loss is minimized compared to open systems which is key when growing in arid climates. Growth of *Chlorella* sp on secondary and tertiary wastewater has been demonstrated, as well as growth using of flue gas as a carbon source. No observed changes in lipid productivity were found compared to control cultures grown in laboratory media.

474778 BUTANOL PRODUCTION BY *CLOSTRIDIUM ACETOBUTYLICUM* IN A SERIES OF PACKED BED BIOFILM REACTORS

Francesca Raganatia, Dipartimento di Ingegneria Chimica, dei Materiali e della Produzione Industriali - Università degli Studi di Napoli Federico II, Napoli, Italy
Alessandra Procentese, Dipartimento di Ingegneria Chimica, dei Materiali e della Produzione Industriali - Università degli Studi di Napoli Federico II, Napoli, Italy
Giuseppe Olivieri, Dipartimento di Ingegneria Chimica, dei Materiali e della Produzione Industriali - Università degli Studi di Napoli Federico II, Napoli, Italy
Maria Elena Russo, Istituto di Ricerche sulla Combustione, Consiglio Nazionale delle Ricerche, Napoli, Italy
Peter Gotz, Department of Life Sciences and Technology / Bioprocess Engineering, Beuth University of Applied Sciences Berlin, Berlin, Germany
Piero Salatino, Dipartimento di Ingegneria Chimica, dei Materiali e della Produzione Industriali - Università degli Studi di Napoli Federico II, Napoli, Italy
Antonio Marzocchella, Dipartimento di Ingegneria Chimica, dei Materiali e della Produzione Industriali - Università degli Studi di Napoli Federico II, Napoli, Italy

The reactor design plays a key role in the fermentative production of biobutanol. The high cell concentration that may be reached in confined – biofilm, membrane, recycling - cell reactors offers high conversion rates. The concentration of solvents in the broth from biofilm reactors reported in literature is not particularly high and it negatively affects the successive stages for butanol recovery (Qureshi et al., 2005). The low concentration of solvents in the produced stream is typically due to the inhibitory effect of solvents on the fermentation. Therefore, the butanol bioreactor productivity is as low as the bioreactor behaviour approaches the CSTR limit.

Another major element of the research on ABE fermentation is the mathematical modeling of this process. The experimental investigations in any typical fermentation process are not only time consuming due to slow kinetics of the process, but also involve significant cost. The performance of ABE fermentation process, in terms of selectivity and yield of solvents, is dependent on many physical and design parameters, in addition to the *Clostridium* strain involved and basic mode of fermentation operation. Thus, optimization of the process involves intense experimentation with different combinations of physical and design parameters, which is rather impractical. Simulation of the fermentation process with mathematical models gives an insight into the characteristics of the process, and also helps to identify the influence of each variable on the overall process.

This contribution reports recent advance on ABE fermentation at Napoli. The study regarded an innovative immobilized cell reactor system: an experimental campaign and the development of a model.

Clostridium acetobutylicum was adopted for the fermentation process. The conversion was carried out in 4 packed bed reactors connected in series: the first reactor of the series was kept under acidogenesis and the successive reactors were kept under solventogenesis. Tests were carried out feeding the reactor system with solutions bearing both glucose as carbon source.

A mathematical model of the PBBRs system was also formulated using kinetic data assessed experimentally. The proposed model was an unstructured-unsegregated model and summarizes biochemical as well as physiological aspects of growth and metabolite synthesis by the production strain.

Introduction

The development of biotechnological processes to produce butanol from renewable resources as eco-sustainable alternatives to the petrochemical production routes (Kumar et al., 2012) is still an open challenge. Acetone-Butanol-Ethanol (ABE) fermentation by clostridia is drawing new interest as a way to turn renewable resources into valuable base chemicals and liquid fuels. Nevertheless the high industrial potential interest for the butanol production by the biotechnological route, some features of the ABE fermentation process hinder its success on an industrial scale. Indeed, the ABE fermentation is characterized by low yield, the acid-solvent two phase feature, and low concentration of butanol due to its inhibiting effect on fermentation. Moreover, the cost of traditional feedstocks - starch and molasses – is a larger fraction of the butanol sale price. Therefore, the use of a low-cost renewable resource is a pre-requisite for the industrial success of biotechnological process to produce butanol. Key issues for the success of the ABE fermentation process include the detailed metabolic analysis of the clostridia coupled with the development of improved engineered strains too (Papoutsakis 2008).

Reactor design and operating conditions play a key role in fermentative productions. The main factors that hinder the commercial development of the traditional ABE batch fermentation processes include low cell density, cell – and relative feedstock - lost at the end of the fermentation, low reactor productivity, high down-times, nutritional limitations and severe product inhibition (Chen and Blaschek, 1999). Process intensification may be obtained by increasing cell concentration in the reactor: cell immobilized reactors and retention membrane reactors are two potential solutions (Qureshi et al., 2005). These reactors configuration takes also the advantage to be operated continuously (Qureshi et al., 2000; Procentese et al., 2015). Continuous bioconversions are characterized by several advantages with respect to batch cultures in biofilm reactors (Qureshi et al., 2000). The main advantages are: the high cell concentration, the reactor operation at high dilution rates – high throughout - without cell washout, high yield, high butanol concentration that enhance butanol recovery performances. Moreover, the biofilm support can be reused.

The reactor design and its optimization for ABE fermentation may take advantages form reactor modelling The performance of ABE fermentation process, in terms of selectivity and yield of solvents, depends on several physical and design parameters, on the Clostridium strain, and on the mode of fermentation operation. Simulation of the fermentation process with mathematical models gives an insight into the characteristics of the process, and also helps to identify the influence of each variable on the overall process.

This contribution reports recent advance on ABE fermentation at Napoli. The study regarded an innovative immobilized cell reactor system: an experimental campaign and the development of a model.

Apparatus and model

The anaerobic solventogenic commercial *Clostridium acetobutylicum* DSM 792 was used for the fermentation process. The conversion was carried out in 4 packed bed biofilm reactors (PBBRs) connected in series: the first reactor (fed with stream bearing the the carbon source) was operated under acidogenesis conditions, and the three successive reactors were operated under solventogenesis conditions. The two phases of the ABE fermentation were operated in separate vessels: acids were produced in first section of the system, the produced acids and the residual sugar were converted in solvents in the second section. The PBBR system performance was characterized in terms of final butanol concentration and productivity as a function of the dilution rate.

The apparatus was made of reactor system, liquid pumps, heating apparatus, device for pH control, and on-line diagnostics (sketch in Fig. 1). The reactor system was made of four fixed beds. Each bed was at the bottom of a 100 mL glass lined pipe (4 cm ID, 8 cm high) jacketed for the heat exchange. Water from an external circulating water bath (Julabo heating circulator MA4) was fed into the jacket of each reactor to keep the operating temperature at 37 °C. The liquid head was controlled by the overflow duct in each reactor: the working volume of each reactor was set at 40 mL. Nitrogen was sparged at the bottom of each reactor to ensure anaerobic conditions. The pH control device – one for each reactor - included a pH-meter, a peristaltic pump, a vessel with NaOH 0.3 M solution and a controller.

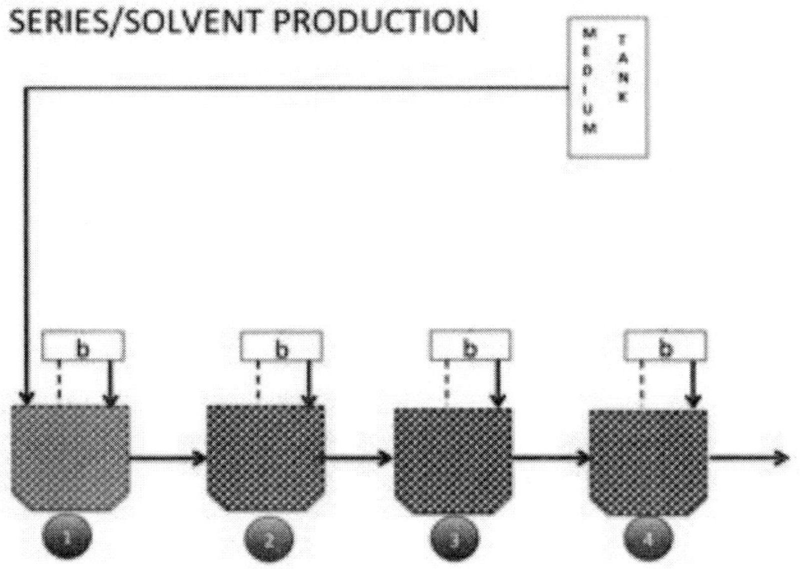

Fig.1: Outline of the apparatus used for the continuous process: A) PBBRs connected in parallel during the start-up phase; B) PBBRs connected in series during the butanol production phase. b: pH measure/control device.

Tygon rings (3/1 mm OD/ID) were chosen as biofilm carriers.
Clostridium acetobutylicum DSM 792 was used. Details regarding the reactivation and pre-inoculum procedures are reported in Raganati et al. 2013.

The composition of the medium fed to the PBBR system is reported in (Procentese et al. 2015a). Glucose (100 g/L concentration) was used as carbon source.

A mathematical model of the PBBRs system was formulated using glucose kinetic data assessed experimentally according to the procedure proposed by Procentese et al. (2015a; 2015b). The proposed model was an unstructured-unsegregated model and summarizes biochemical as well as physiological issues of growth and metabolite synthesis by the production strain. The key fermentation rates were expressed and evaluated with regard to substrate consumption and butanol end-product inhibitory effects.

Kinetic data regarding the growth rate (acidogenesis) and the butanol production rate (solventogenesis) were from specific experimental campaigns (details in Procentese et al. 2015a and 2015b). The main assumptions of the proposed model are reported hereinafter (Raganati et al., 2016).

- the PBBR system was assumed as a series of CSTR;
- the biomass present in each PBBR as free cells and immobilized cells was a heterogeneous cell population consisting of: acidogenic cells, solventogenic cells and spores ;
- the kinetics of cell growth and butanol production of biofilm-cells were assumed equal to those of the free cells;
- cells attachment and detachment processes were considered.

Experimental results

Two PBBR configurations were used (Raganati et al., 2016).

Parallel configuration) The four fixed beds were operated in parallel. This configuration was used during start-up to promote biofilm formation.

Series configuration) The four fixed beds were connected in series (Fig. 1). This configuration was used during butanol production after a biofilm layer had formed in each unit in parallel mode.

Butanol production tests were carried out with the 4 PBBRs connected in series (Fig.1) and operated at pre-set conditions. The pH of the reactor 1 (Fig.1) was set at 5.5 to promote acidogenesis conditions. The pH of the reactors 2, 3, and 4 (Fig.1) was set at 4.7 to promote solventogenesis conditions.

The overall dilution rate (D) – ratio between the feeding flow rate and the total volume of the 4 fixed beds – ranged between 0.05 and 1.4 h^{-1}. After setting the dilution rate, the reactor system was operated until steady state conditions were reached: metabolite and glucose concentration in each reactor staying constant for at least 5 times the space-time of the reactor.

Fig. 2 reports the concentration of butanol and glucose in the four PBBRs measured under steady state conditions for D=0.15 h^{-1}. It is interesting to analyse the substrate and metabolite concentration in the four reactors from the PFR behaviour point of view. The gradual decrease in glucose concentration (Fig. 2B), and the gradual increase in solvent concentration (Fig. 2A) along the PBBR series reproduce the typical behaviour of batch fermentation. As expected, the formation of a stable biofilm in the four units and the choice of the proper operating conditions lead to a successful reactor system that behaves like a PFR with continuous inoculum and constant biofilm concentration.

The PBBRs system was successfully operated for more than three months to produce butanol. The reactor performances – butanol productivity and butanol concentration – were assessed as a function of the reactor system dilution rate. By tuning the D it was possible: i) to totally convert the carbon source (D <0.15 h^{-1}); ii) to maximize the concentration of butanol in the produced stream (about 15 g/L at D = 0.65 h^{-1}); iii) to maximize the butanol productivity (about 12.6 g/Lh at D = 0.9 h^{-1}).

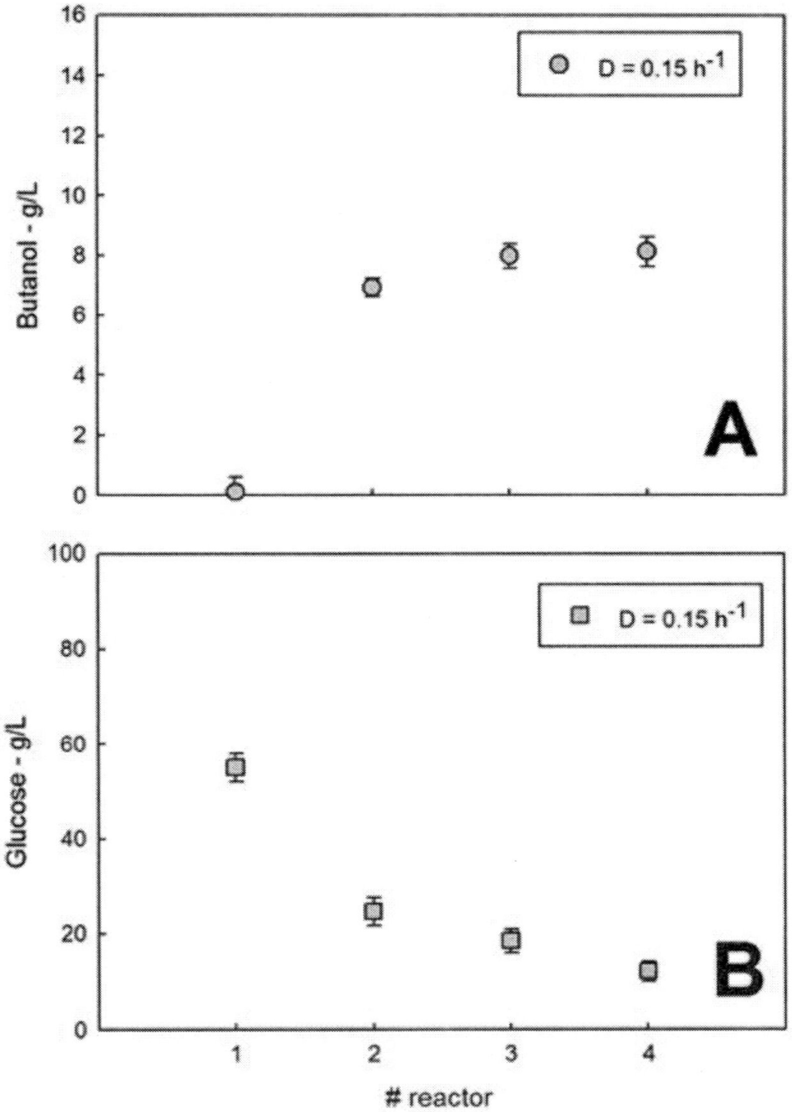

Fig.2: Butanol and glucose concentration measured along the PBBR series at D=0.15 h^{-1}.

Model results

Fig. 3 reports the estimated concentration of butanol and glucose in the four PBBRs for D=0.15 h^{-1}. It is evident that the developed model succeeded in reproducing the behavior of the PBBRs system.

Acknowledgements

The authors thank the European Union's Horizon 2020 research and innovation program for their financial support to the project Waste2Fuels (grant agreement No 654623).

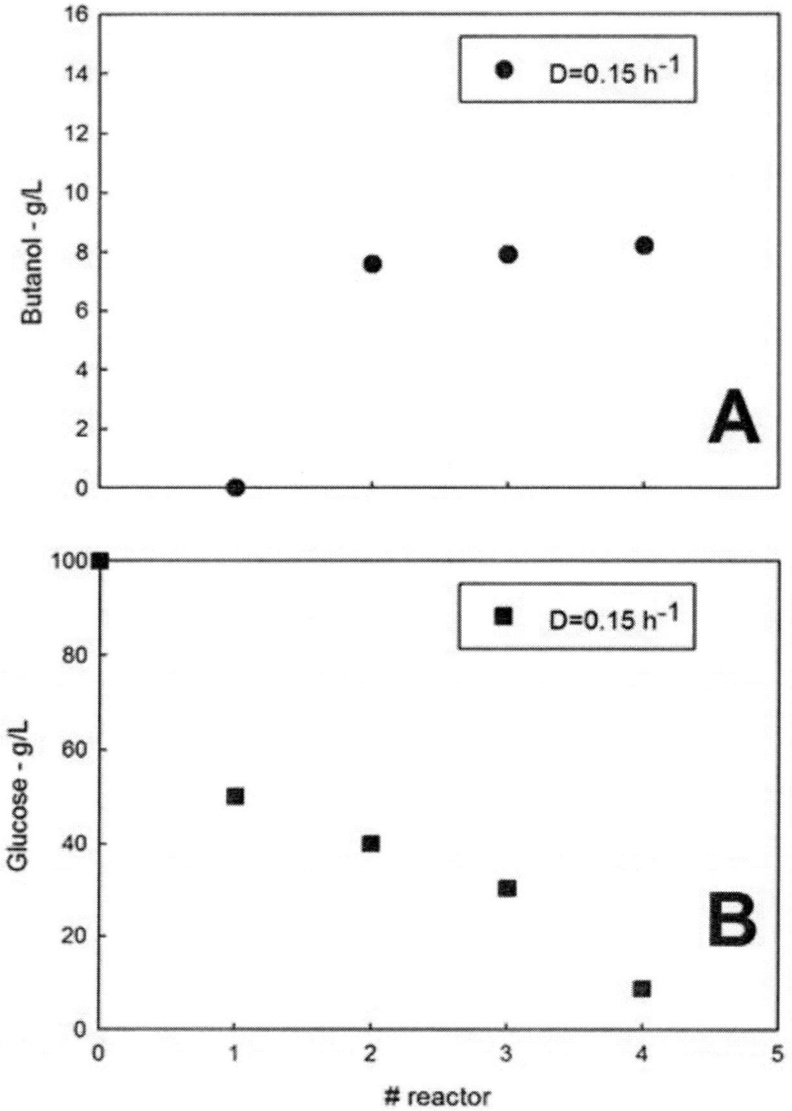

Fig.3: Estimated butanol and glucose concentration along the PBBR series at D=0.15 h^{-1}.

References

1. Chen CK, Blaschek HP. Acetate enhances solvent production and prevents degeneration in *Clostridium beijerinckii* BA101. Appl Microbiol Biotechnol 1999; 52:170–173
2. Kumar M, Goyal Y, Sarkar A, Gayen K. Comparative economic assessment of ABE fermentation based on cellulosic and non-cellulosic feedstocks. Appl Energ 2012; 93:193–204.
3. Papoutsakis, E. T. Engineering solventogenic clostridia. Current Opinion in Biotechnology 2008 19(5), 420–9.

4. Procentese, A., Raganati, F., Olivieri, G., Russo, M. E., Salatino, P., Marzocchella, A. Continuous xylose fermentation by *Clostridium acetobutylicum* – Kinetics and energetics issues under acidogenesis conditions. Bioresource Technol 2015a 164, 155–161.
5. Procentese A, Raganati F, Olivieri G, Russo ME, Salatino P, Marzocchella A. Continuous lactose fermentation by *Clostridium acetobutylicum* – Assessment of solventogenic kinetics. Bioresour Technol 2015b; 180:330–337.
6. Qureshi N, Schripsema J, Lienhardt J, Blaschek HP. Continuous solvent production by *Clostridium beijerinckii* BA101 immobilized by adsorption onto brick. World J Microb Biot 2000; 16:337-382.
7. Qureshi, N., Annous, B. A., Ezeji, T. C., Karcher, P., & Maddox, I. S. (2005). Biofilm reactors for industrial bioconversion processes : employing potential of enhanced reaction rates. Microbial Cell Factories, 21, 1–21.
8. Raganati, F., Olivieri, G., Procentese, A., Russo, M. E., Salatino, P., Marzocchella, A. Butanol production by bioconversion of cheese whey in a continuous packed bed reactor. Bioresour Technol 2013; 138:259–265.
9. Raganati, F., Procentese, A., Olivieri, G., Russo, M. E., Götz, P., Salatino, P., Marzocchella, A. Butanol production by *Clostridium acetobutylicum* in a series of packed bed biofilm reactors. Chemical Engineering Science 2016; 152; 678–688.

466294 Effect of Environmental Factors Towards Patchoulol Production By *Chlamydomonas Reinhardtii*

Abstract:

Patchoulol ($C_{15}H_{26}O$) is a sesquiterpene alcohol and a valuable compound in the fragrance and pharmaceutical industry. Traditionally, patchouli oil is extracted from the leaves of a plant, *Pogostemon cablin*. This approach faces the threat of limited and intermittent supply as these plants only grow in select regions of the world. But now with the advance of genetic engineering, it is possible to genetically modify microorganisms such as yeasts and bacteria to produce patchoulol. Photosynthetic microorganisms, such as microalgae in particular, would be of high interest as they have the advantage of directly converting carbon dioxide to high-value products through carbon fixation and photosynthesis. In this work, we have cultivated a patchoulol-producing strain of *Chlamydomonas reinhardtii* and elucidated the effect of light and temperature towards cell growth and patchoulol production. We also considered the effect of light-and-dark cycles towards patchoulol productivity. The findings from these environmental factors not only increase our understanding of patchoulol expression in microalgae, but also represent a step towards advancing this production platform to other target molecules. It is envisioned that this platform would be applicable to other terpenoids such as bisabolene and limonene, both hydrocarbon precursors to a range of industrially relevant chemicals.

470048 Nutrient Capture By a Sustainable Symbiotic Biofilm: Simultaneous Phosphorous and Nitrogen Recovery By Attached Fungi and Microalgae Biofilm

Abstract:

Nutrient pollution, a condition of excess phosphate (P) and nitrate (N) from point and non-point sources, is a major water quality concern worldwide due to the tremendous volume of urban, agricultural and industrial wastewater production [1]. The elimination of phosphorus from effluents before discharging is made mandatory as even traces of P (above 0.02 mg/L) can cause eutrophication, also P level is regulated to be lower than 0.05 mg/L, if streams discharge into lakes or reservoirs. On the other hand, the P reserves are being depleted swiftly which apparently is linked to global food security, with more than 90% of phosphorus used for agricultural purpose [2]. Currently a curative solution to combat water pollution by recuperating N and P, recycling the P for food production by closing the nutrient loop to effectively manage dwindling rock P reserve are the major challenges. The most common phosphorus removal techniques for municipal and industrial wastewater are biological treatment and chemical precipitation. Bio-valorization offers an excellent alternative to chemical treatment technologies as the biomass produced can be a source of protein and other valuable bio-chemicals also it is an eco-friendly and low cost technology.

In this present work, a novel composite biofilm called mycoalgae biofilm [3] was developed using the isolated strains of polyphosphate accumulating fungi and microalgae and we want to apply this platform for the efficient recovery and recycling of nutrients in waste stream. The effect of various process conditions (fungal strain, carbon levels, temperature, P/N ratio) on the nutrient recovery efficiency and the changes in the biofilm biomass composition was studied. Since the biofilm will be used for nutrient recovery from an array of samples a synthetic medium for the cell growth with the initial P (40 mg/L), N (200 mg/L) and other nutrients were used as the medium for the biofilm development. Depending on the nature of the waste water and its composition, and the fungi-microalgae combination intended to use for bio-remediation, the developed mycoalgae biofilm can be a sustainable microbial process for food, feed or energy source.

Methods: Water and soil samples were collected from Sarita wetland near UMN to isolate the native fungi for the biofilm formation. Fungal isolates were identified by genetic identification of fungal internal transcribed spacer (ITS) regions. Maximum Phosphorous and Nitrogen removal was observed in strain Sa7 (*Mucor sp.)* and *Chlorella sp,* combination which was considered for the biofilm formation and Nutrient recuperation. The morphology of the biofilm was also studied using CLSM and SEM at the optimized conditions to study the interaction between the cells. The nutrient removal efficiencies were calculated using the equation: RE_i = (S_{i0} â€“ S_{it})/S_{i0}; where RE_i is the removal efficiency of the nutrients *i* (TN, TP, PO_4-P) or the pollutants (*i*) using the mycoalgae lichen type biofilm; S_{i0} and S_{it} are the initial and final concentration of component *i* after *t* days. The rate of nutrient and pollutant removal by mycoalgae biofilm is an important factor which will be calculated by; dS/dt = r_i = (S_{i0} â€“ S_{it})/ (t_i â€“ t_0); Where: r_i (g/L/d) is the removal rate of substrate *i* (TN, TP, PO_4-P).

Conclusion: The composite biofilm shows promising results in waste water treatment especially in removing the N, P, suspended solids and other organics from nutrient polluted water in a single-step reactor with easy biomass recovery and high efficiency, when compared to the individual pure cell cultures. The biofilm attaches to the polymer matrix completely leaving behind the clean water for recirculation, and the composite biomass for nutrient recycle. The biofilm composition can also be tailored based on the influent stream components, feasibility of the strains to grow together and its nutrition value for further use. The commercial expansion of this biofilm system could be a step ahead in having an integrated and sustainable water treatment systems. **References**

1. E.K. Wilson, Danger from microcystins in Toledo water unclear, Chem. Eng. News, 92 (32) (2014), p. 9
2. Ye, Y., Gan, J., & Hu, B. (2015). Screening of Phosphorus-Accumulating Fungi and Their Potential for Phosphorus Removal from Waste Streams. Applied biochemistry and biotechnology, 177(5), 1127-1136.
3. Bo Hu, Aravindan Rajendran, A Novel Microalgae Cultivation Technology Using Fungi and Lichen Biofilm, UMN invention disclosure case # 20140274

461292 Incorporating Hydrodynamics into Spatiotemporal Metabolic Models of Bubble Column Gas Fermentation

Abstract:

A promising route to renewable liquid fuels and chemicals is the fermentation of carbon monoxide containing gas streams to synthesize desired products such as ethanol and 2,3-butanediol. We recently developed a spatiotemporal metabolic model for bubble column reactors with the CO fermenting bacterium *Clostridium ljungdahlii* as the microbial catalyst [1]. This first generation model was based on the simplifying assumptions of ideal plug flow for the vapor phase and plug flow plus axial dispersion for the liquid phase. However, industrial bubble column reactors are complex multiphase processes in which spatial variations in the gas superficial velocity, volumetric holdup and interfacial area can have profound e ects on fermentation performance.

We discuss the development of a second-generation bubble column model that accounts for the e ects of gas phase hydrodynamics on cellular growth, nutrient update and byproduct synthesis rates in a temporarily and spatially resolved manner. The modeling approach involves the coupling of a genome-scale metabolic reconstruction with mass and momentum balances that allow the calculation of local gas metabolite concentrations, superficial velocity, volumetric holdup and interfacial area as well as local concentrations of biomass and liquid phase metabolites. The resulting set of partial di erential equations with embedded linear programs is spatially discretized and numerically integrated using the MATLAB based code DFBAlab. Simulations results are shown for two CO fermentating organisms: the research model *C. ljungdahlii* and the industrially relevant *Clostridium autoethanogenum*. Model predictions are compared to those generated without gas phase hydrodynamics and to experimental data collected from a laboratory scale bubble column reactor with respect to CO conversion, biomass production, ethanol titer and ethanol-to-acetate ratio.

1. Chen, J., J. A. Gomez, K. Ho ner, P. I. Barton and M. A. Henson, "Metabolic Modeling of Synthesis Gas Fermentation in Bubble Column Reactors," Biotechnology for Biofuels, 8, 89, doi:10.1186/s13068-015-0272-5 (2015).

464191 A Commercial Demonstration of Biorefinery of Lipids--Coproduction of Biodiesel and 1,3-Propanediol

Abstract:

Lipase-catalyzed transesterification (also esterification) from oils or fats for biodiesel production has some advantages over chemical-catalyzed transesterification, such as environmental friendship, lower energy consumptionï¼Œwidely suitable for low quality vegetable oils. However, the low stability of the lipase due to alcohol has been being regarded as the main hurdle to the industrialization of lipase-catalyzed biodiesel production. Tsinghua University has developed a novel process to eliminate the inhibition of methanol on lipase, thus the operational life of the lipase could be improved into more than 300 cycles from several (1-5) cycles while traditional process applied. The patented novel process was successfully demonstrated in a pilot plant (200kg/day) and commercial plant (50,000ton/year) in China. It has also been being demonstrated in a pilot-plan in Brazil.

As a by-product, glycerol will be produced at about 10% of biodiesel during the process of biodiesel production. More and more concerns were pay on how to utilize crude glycerol. It could be a promising way to produce 1,3-propanediol (PDO) by fermentation with crude glycerol as feedstock. PDO is a valuable monomer to synthesize polytrimethylene terephthalate (PTT). PTT has excellent properties for textile fibre or plastic, and it was found to have some better properties than other polyesters such as PET, PBT. Tsinghua University has developed a novel flexible process for PDO production by fermentation from glycerol or glucose, and it was successfully commercialized with capacity of 20,000 tons/year PDO .

471918 Ecosystems As Unit Operations: Designing Integrated Networks of Technological and Ecological Systems

Abstract:

One of the requirements to meet the goal of sustainable design is to ensure that impacts of engineering systems do not exceed the regenerative capacity of ecological systems [1], in order to prevent unintended harm. Industries striving towards sustainable development also aim to become carbon and emissions neutral, with zero water withdrawal. Many companies aim to go even beyond such goals by having "net positive" manufacturing systems. While the conventional approach to meeting this goal has been to increase the use of renewable fuels, optimize the system efficiency to reduce emissions and rely on end-of-pipe pollution control methods, such solutions tend to neglect the ecological overshoot resulting in some perverse outcomes.

In this work, we present an approach to designing coupled techno-ecological systems in order to minimize ecological overshoot within an "industrial-serviceshed", by including ecological systems in a manner analogous to unit operations within the engineering design framework. As a result, simultaneous decisions are made regarding the design of engineering systems to minimize the ecosystem service demand and the design of supporting ecological systems to supply the necessary ecosystems services demanded by industrial sites. Sustainability indices for such systems are quantified using the TES (Techno-Ecological Synergy) framework [1], as a function of the design variables and the objective is to identify optimal coupling between Technological and Ecological systems, to minimize the overshoot.

Traditional engineering unit-operation models capture the demand for ecosystem service, while the supply is quantified by spatially explicit unit-operation models of ecosystems. Similar to process level constraints regarding the plant operating conditions, feedstock availability etc. ecological constraints like land availability, type of ecological systems and other geo-spatial constraints based on remote-sensing and aerial survey are identified and included within the design problem. The objective of TES system design is to maximize the net ecosystem service supply, quantified by the ecosystem service supply less the demand as opposed to the conventional approach of minimizing only the ecosystem service demand.

The TES design framework is applied to the design of a biofuel production system. Ecological systems included within the design framework include a forest ecosystem captured using the Urban Forest Effects model [2] and a horizontal subsurface flow wetland systems for treating the wastewater, designed using steady-state first order equations of an ideal plug flow reactor [3]. The problem is formulated as a multi-objective nonlinear program (Mo-NLP) and tradeoffs are identified between maximizing the net present value of the system and the net ecosystem service demand for multiple ecosystem services. The dynamic behavior of ecological systems and changes in the design and operation of systems are captured using a multi-period approach. Seasonal variations due to weather fluctuations are also bound to cause a change in ecological processes and thus robust designs are identified at each time-period accounting for these external disturbances. The problem formulation also accounts for the uncertainty associated with the critical design and operating variables.

Preliminary results from the case study demonstrate that systems designed by accounting for ecosystem service supply results in novel designs that have a higher economic value, and higher sustainability index, resulting in "win-win" designs. Thus, coupled TES systems are expected to have economic and environmental advantages over conventional systems design, mainly due to the reduction in external inputs by moving towards a closed loop production system.

References

[1] Bakshi, Bhavik R., Guy Ziv, and Michael D. Lepech. "Techno-ecological synergy: A framework for sustainable engineering." Environmental science & technology 49.3 (2015): 1752-1760.

[2] Nowak, David J., and Daniel E. Crane. "The Urban Forest Effects (UFORE) Model: quantifying urban forest structure and functions." (2000).

[3] Rousseau, Diederik PL, Peter A. Vanrolleghem, and Niels De Pauw. "Model-based design of horizontal subsurface flow constructed treatment wetlands: a review." Water research 38.6 (2004): 1484-1493.

459809 Economic, Environmental and Social Cost Optimization of Biomass Supply Chain for Electricity Generation in WI, USA

Abstract:

There is an increasing interest in identifying the optimal biomass supply chain for the renewable energy system, due to relatively low energy density and high procurement cost of biomass feedstock. The goal of this study is to develop an integrated framework to determine the optimal supply chain network for bioelectricity generation in Wisconsin, USA. First, forest residues collection sites, road and rail network across WI, MN and upper MI were mapped in Arc-GIS. Next, the optimal location of co-generation facility was identified by screening all potential sites by government policy, local community interest, land availability, distance to transmission lines and road/rail network. Finally, biomass supply chain was optimized by minimizing the economic, environmental and social cost of shipping forest residues from forest to the facility. Upstream (cradle-to-gate) GHG and criteria pollutants emissions of feedstock were also quantified for all potential facility sites. The result shows that Superior WI is the best location to build a bioelectricity facility in terms of economic, environmental, and social sustainability. In addition, the quantities, modes and distances of forest residues transportation were determined in order to meet the 0.5 million ton (current capacity) and 1 million ton (future capacity) demand of the bioelectricity facility.

472967 Renewable/Fossil Fuel Supply Selection and Supply Chain Optimization for a Synergized Upstream, Downstream and Power Economy System

Abstract:

The current environment of low oil prices constitutes a substantial headwind for much of the energy value chain, especially for energy systems where the upstream, downstream, and power sectors are synergized with the aid of governmental subsidies. In such cases, the economic viability of the downstream and power sectors depends heavily on the technology costs and the government regulation on oil prices, which are often fixed. In the current decade, energy systems with an abundant supply of oil were able to achieve sizable margins in the downstream sector within the high oil price environment. However, the uncertainty of future oil prices makes it difficult to put forward strategies for the energy and petrochemical sectors. An energy model that takes into account the incorporation of renewable resources mixed with fossil fuels, a dynamic selection of hydrocarbon prices, and a flexible downstream sector that is capable of liquid or gas feedstock inlets can lessen the burden that oil price fluctuations have. In this work, we present a model where static and dynamic optimization techniques are applied to the dynamic identification of prices that optimize supply decisions and minimize fuel and technology costs within a synergized upstream, downstream and power economy system. Emphasis is placed on understanding the role of modeling in the energy sector and understanding energy balances to develop a modern integrated energy planning model.

471652 Synthesis of Renewable Energy Supply Networks Considering Different Frequencies of Fluctuations in Supply and Demand

Abstract:

The demand for and dependency on fossil resources, coupled with growing environmental consciousness and new economic opportunities and developing/utilising clean and renewable energy sources on a larger scale, are among the major challenges and priorities of our society. In 2015 at least 164 countries had renewable energy targets, and most also had renewable energy support policies (Sawin et al., 2015). Among the most promising technologies to help achieve these goals are those using solar irradiance, wind and biomass (Davis and Martín, 2014).

In this study, a biorefinery supply network integrating wind and solar energy (Čuček et al., 2015), involving several biomass and waste sources for the production of biofuels and food (Čuček et al., 2014), is further extended by: i) applying more detailed models for wind and solar energy with more realistic data, and ii) better accounting for intermittent renewable energy sources, especially of wind and solar energy, and electricity demands. The dynamic model corresponds to a multi-period model, and considers different frequencies in supply and demand. In those cases where there is a more pronounced dynamic, the time interval selected is shorter. For supply and demand with longer-term fluctuations, the time interval is longer. Different time periods are applied leading to supply network models with a large number of variables and constraints, and to long computational times. For biomass and biofuels the time periods are longer, while for solar, wind and electricity consumption, the time intervals are shorter.

The proposed model is demonstrated on a case study of a relatively small region in Europe. Several scenarios are considered regarding possible percentage share of renewables in the future energy mix, preferably reaching 100% of renewable energy systems. Furthermore, a sensitivity analysis is performed regarding the availability of energy sources, demand, and prices of biomass, energy technologies and products.

Acknowledgement

The authors acknowledge the financial support of the Slovenian Research Agency (programs P2-0032 and P2-0377), from a bilateral project between Slovenia and the United States entitled "Development of Methodology for the Synthesis of (Bio)chemical Supply Chains," from the Ministry of Science, Education and Sports of the Republic of Croatia and the EC under project CARBEN (NEWFELPRO Grant Agreement No. 39), and from a bilateral project between Slovenia and Croatia entitled SLOCRORES, "Planning of 100% RES communities by use and combination of Total Site Integration and Renewislands methodologies."

References

Čuček, L., Martín, M., Grossmann, I.E., Kravanja, Z., 2014. Multi-period Synthesis of Optimally-Integrated Biomass and Bioenergy Supply Network. Computers and Chemical Engineering 66, 57-70.

Čuček, L., Martín, M., Kravanja, Z., 2015. Integration of Wind and Solar Energy within Continental Biorefinery Supply Network. Proceedings of 2015 AIChE Annual Meeting, November 8 – 13, 2015, Salt Lake City, UT, USA.

Davis, W., Martín, M., 2014. Optimal year-round operation for methane production from CO_2 and water using wind and/or solar energy. Journal of Cleaner Production 80, 252-261.

Sawin, J.L., Sverrisson, F., Rickerson, W., Lins, C., Musolino, E., Petrichenko, K., Rickerson, W., Sawin, J., L., Seyboth, K., Skeen, J., Sovacool, B., Williamson, L.E., 2015. Renewables 2015 - Global status report, Key findings 2015, Paris, France.

469724 A Holistic Look to Carbon Capture and Use in Sustainable and Economically Enhanced Bio-Based Supply Chains

Abstract:

Carbon capture and storage is considered worldwide as an alternative to take into account for cost-effective mitigation of CO_2 emissions. Notwithstanding, recent years have shown significant drawbacks associated with carbon capture and storage (CCS) to secure the CO_2 geological storage from industrial sources or power plants. Geological storage faces the possibility of leakage, liability issues, long-term problems with public acceptance of storage sites on land, and limited cost-effective storage capacity in some key regions (Cuéllar and Azapagic, 2015). Otherwise, sustainability concerns, evidence that a more holistic view is necessary. This includes the whole energy supply chain and a deeper insight on the protagonist role of carbon capture and utilization (CCU), specifically for the chemical industry, and as a part of the share of the environmentally friendly technological options of the future.

CCU provides a CO_2 management solution for a limited amount of CO_2. For instance, the CO_2 generated by a single coal power plant could potentially supply much of the CO_2 demand for CO_2-based methanol plants that supply a growing demand in Europe. This would avoid the use of fossil fuel, like natural gas, which is used nowadays in the conventional synthesis of methanol (Pérez-Fortes et al., 2016).

This work combines a bio-based energy supply chain with CO_2 capture and utilization, for economic optimisation based on a CO_2 emissions constraint. The problem is formulated as a mixed-integer linear program in GAMS. This study presents a complete and flexible supply chain superstructure, as well as the supporting framework developed to answer pending issues on the actual economic and environmental benefits achievable in this complex scenario for both, the biomass-coal plant and the CO_2 utilization plant; i.e. location of intermediate steps, fraction of CO_2 emissions captured per plant, or size of the CO_2 utilization plants, among others. Moreover, eventual incentives and environmental revenues are discussed, to make an economically viable business case. A large-size, real case study is used in this analysis. Two key scenarios are taken into account: (i) No use of biomass, neither capture nor utilization of CO_2 is contemplated; (ii) Biomass, capture and CO_2utilization are all contemplated. Finally, conclusions and guidelines are given.

References

Cuéllar-Franca R, Azapagic A. Carbon capture, storage and utilisation technologies: A critical analysis and comparison of their life cycle environmental impacts. *Journal of CO2 Utilization*. Elsevier. Volume 9, March 2015, Pages 82â€"102.

Available from: doi:10.1016/j.jcou.2014.12.001

Pérez-Fortes M, Laínez-aguirre JM, Puigjaner L. Economic and environmental assessment of CO_2 capture and utilisation technologies in bio-based supply chains. *Computers and Chemical Engineering*. 2016; in press.

.

459900 Optimal Integration of Renewable Based Processes for Fuels and Power Production: Case Study in Spain

Abstract:

Although renewable resources are plentiful, their sources are scattered. Therefore, to meet a certain demand of power and fuels using renewable sources, and mitigate the variability in their availability, process integration at local or greater scale has become an interesting engineering and technical challenge. Weekman (2010) and Yaun and Chen (2012) presented overviews regarding the integration possibilities as a perspective for the future combination of different sources of energy. In order to help with those decisions, process system engineering has the tools to compare sources and allocations in search for the optimal options at local and wider scales. Most of the studies either focus on fuels supply (Cucek et al 2014, Elia et al 2012) or power supply based on the unit commitment problem (Saravanan, et al 2013). However, power is most of the times produced by using a fuel, which is a chemical that can also be synthesized. Therefore, both supply networks are linked and must be addressed simultaneously. Over the last years plant size integration of renewable resources has been considered (Vidal and Martín, 2015, Martín and Davis, 2016, Martín and Grossmann, 2016). However, the integration at a larger scale, beyond the plant level, has not been attempted.

In this work we have developed a framework for the optimal integration of renewable sources of energy to produce fuels and power. A network is developed using surrogate models for various technologies that use solar energy, PV solar, CSP or algae to produce oil, wind technology, biomass based syngas to ethanol, methanol, FT-liquids and thermal energy, hydroelectric power and waste based power plant via biogas production. Hydrogen can be produced if there is a surplus of energy. It can be stored, by producing methanol or methane. Methane can be further used to produce power at need, while methanol is employed for oil transesterification to produced biodiesel. In both cases, the carbon source is CO_2, that has been captured during the gasification based processes or it imported from outside the network. The corresponding processes are modeled using an input- output approach based on the results of detailed optimization studies previously developed by the authors. However, we do not model an entire process from biomass to fuels as a black box, but we break it into sections that produce intermediates that can be used for a different purpose, i.e. syngas can be used to produced ethanol FT-liquids, etc, or because various technologies are available for that; i.e. dry or wet cooling for power plants. The model is formulated as an MILP that allows determining the optimal selection of technologies to meet certain demand. Both cost and environmental objective functions are considered.

We apply the network to evaluate process integration at different scales, from region level, where we also evaluate the effect of uncertainty in renewable resources, up to the level of an entire country (Spain). Solutions under uncertainty are more robust and expensive, about 25% higher requiring the use of further technologies to meet the demand. For Spain, and using 1% of the area, it is possible to substitute 20% of fossil fuels for transportation and 100% of the power demand. The solution suggests the use of Hydropower, and oil production in all regions, while bioethanol biodiesel plants are allocated close to demand points such as large population areas. Up to 44% of fuels and total power can be substituted with the current technology development status. However, we can only reach 90% substitution by using 20% area with the availability and efficiency of current processes. The investment required for this option is more than twice that to reach 20% fossil fuels substitution.

References

Cucek, L., MartÃn, M., Grossmann, I.E., Kravanja, Z (2014) Large-Scale Biorefinery Supply Network â€" Case Study of the European Union. Comp. Aid. Chem. Eng. Vol33 , 319-324.

Elia, J.A., Baliban R.C., Floudas, C.A. (2012) Nationwide energy supply chain analysis for hybrid feedstoch processes with significant CO2 emissions reduction. AIChE J. 58 (7) 2142-2154.

Martín, M., Grossmann, I.E. (2016) Optimal integration of a self sustained algae based facility with solar and/or wind energy. submitted *Energy*.

Martín, M, Davis, W. (2015) Integration of wind, solar and biomass over a year for the constant production of CH_4 from CO_2 and water Comp Chem Eng 84, 314-325,

Saravanan, B., Da, S., Sikri, S., Kpothari, D.P. (2013) A solution to the unit commitment problem-a review. Frointiers in Energy. 7(2) 223-236

Vidal, M., Martín, M.(2015) Optimal coupling of biomass and solar energy for the production of electricity and chemicals.Comp. Chem. Eng 72, 273-283

Weekman, V.W., 2010. Gazing into an energy crystal ball. CEP June 2010; 23-27

Yuan, Z., Chen, B., 2012. Process Synthesis for Addressing the Sustainable Energy Systems and Environmental Issues. AIChE J. 58 (11), 3370-3389

454781 *Featured:* **Energy Conversion Considerations in Solar-Driven Photoelectrochemical Water Splitting and Carbon Dioxide Reduction**

Abstract:

A practical method to use sunlight to generate chemical fuels could dramatically change the landscape of global energy generation and storage. One specific implementation of this concept is an integrated photoelectrochemical (PEC) device that operates without added electrical bias with sunlight as the only energy input, producing either hydrogen by water splitting or carbon-based fuels via carbon dioxide reduction. Experimental demonstrations of such systems date back to the 1970s, and, since that time, there have been improvements in both the functional understanding and performance of the devices.

There are a number of design considerations which arise when considering the full solar to fuel conversion system, both for water splitting and carbon dioxide reduction. These include the type and number of semiconductor junctions used to generate the requisite photovoltage, the use of co-catalysts, the electrolyte conditions, a product separation method such as a membrane, and the geometric layout and size of components.

To date, all approaches which have a solar to hydrogen (STH) conversion efficiency of >5% employ either 2 or more semiconductor junctions arranged in a voltage-additive tandem configuration. The degree of integration varies widely, from fully integrated "artificial leaves" to systems with decoupled photovoltaics and catalysts. A small subset of devices employ a mechanism, such as a proton or anion conducting membrane, to separate products and yield a pure H_2 stream. Reported STH conversion efficiencies range from <1% to over 20% [1]. However, there are very few reports of long term operational stability, which is a clear prerequisite for a positive energy return on investment [2].

Electrochemical reduction of CO_2 to fuel molecules such as methanol and ethanol could form the basis for the production of renewable and sustainable transportation fuels, replacing the fossil fuels which are used today. There has been substantial recent progress in solar driven PEC CO2 reduction, with a number of reports of energy conversion efficiencies of over 1%. While these values are larger than those estimated for biomass generation by plants and algae (up to 1%) [3], the natural system utilizes dilute CO_2 in the atmosphere as a feed stock and generates a separated fuel product. Thus, product separation and the selective production fuels remain as outstanding challenges for the PEC technology.

This material is based upon work performed by the Joint Center for Artificial Photosynthesis, a DOE Energy Innovation Hub, supported through the Office of Science of the U.S. Department of Energy under Award Number DE-SC0004993.

[1] Ager, J. W.; Shaner, M. R.; Walczak, K. A.; Sharp, I. D.; Ardo, S. *Energy Environ. Sci.* **2015**, *8* (10), 2811–2824.

[2] Sathre, R.; Scown, C. D.; Morrow, W. R.; Stevens, J. C.; Sharp, I. D.; Ager, J. W.; Walczak, K.; Houle, F. A.; Greenblatt, J. B. *Energy Environ. Sci.* **2014**, *7*, 3264–3278.

[3] Blankenship, R. E.; Tiede, D. M.; Barber, J.; Brudvig, G. W.; Fleming, G.; Ghirardi, M.; Gunner, M. R.; Junge, W.; Kramer, D. M.; Melis, A.; Moore, T. A.; Moser, C. C.; Nocera, D. G.; Nozik, A. J.; Ort, D. R.; Parson, W. W.; Prince, R. C.; Sayre, R. T. *Science.* **2011**, *332*, 805–809.

466023 An Enabling Start-up Support System for CO_2 Utilisation: A Case Study on Barriers to Commercial Success

Abstract:

The topic of this paper is sustainable entrepreneurship focussing on 'emerging Davids' (Hockerts, Wüstenhagen 2010) and their endeavour to bring CO2 utilisation technologies to the market.

The problem is that a proof-of-concept on a technically relevant scale is essential for new technologies in order to attract external stakeholders such as investors, customers and other partners. These proofs of a scale-up production for CO2 utilisation technologies are ranging from demonstrators over pilot production facilities to large commercial plants and are most of the times very capital intensive. This capital intensity often leads to the point that new technology ventures fail to successfully commercialise their products. This results in a lack of CO2 utilisation start-ups and little radical innovation to tackle current societal challenges like resource depletion and climate change. Recent events such as the COP21 in Paris and the G7 summit in 2015 may be promising means in order to boost sustainable-oriented innovation by investing more and more money in clean technologies over the next decades, but an effective allocation and a sophisticated design for a support system are still relatively unexplored.

In order to give managerial recommendation on the design of a favourable support system for capital intensive CO2 utilisation technologies a better understanding on the barriers to those innovation is necessary. A broad portfolio of CO2 utilisation technologies are in scope of this contribution; ranging from chemicals production (e.g. intermediates and specialty chemicals from CO2) over CO2 mineralisation (e.g. building materials from CO2) to CO2 to fuels (Hendriks et al. 2013). This work builds on a multilevel perspective from current barrier research (Hueske, Guenther 2015) to investigate all relevant barriers to commercial success. The authors choose a dynamic capabilities approach (Eisenhardt, Martin 2000) in combination with stakeholder theory (Freeman 2004) to analyse both internal as well as external barriers.

A exploratory multiple case study approach (Eisenhardt, Graebner 2007; Yin 2013; Eisenhardt 1989) was used to investigate the CO2 utilisation start-up landscape worldwide. CO2 utilisation may not only be a door-opener for large emission abatement technologies but is also a potent resource efficiency technology by feeding CO2 back into the carbon-based economic system (Styring et al. 2014; Naims et al. 2015). A total of 20 semi-structured interviews have been carried out, transcribed, coded (Strauss, Corbin 2015), and triangulated with other available data sources during a 5-month-period in 2015/2016.

This work is funded by the EIT Climate-KIC enCO2re programme – 'enabling CO2 re-utilisation'.

References:

Eisenhardt, Kathleen M. (1989): Building theories from case study research. In *Academy of management review* 14 (4), pp. 532–550.

Eisenhardt, Kathleen M.; Graebner, Melissa E. (2007): Theory Building from Cases: Opportunities and Challenges. In *The Academy of Management Journal* 50 (1), pp. 25–32.

Eisenhardt, Kathleen M.; Martin, Jeffrey A. (2000): Dynamic capabilities: What are they? In *Strategic Management Journal;* 21, pp. 1105–1121.

Freeman, R. Edward (2004): The stakeholder approach revisited. In *Zfwu* 5 (3), pp. 228–241.

Hendriks, Chris; Noothout, Paul; Zakkour, Paul; Cook, Greg (2013): Implications of the Reuse of Captured CO2 for European Climate Action Policies. With assistance of Carbon Counts Ltd. ECOFYS Netherlands B.V. Utrecht. Available online at http://www.scotproject.org/sites/default/files/Carbon%20Count,%20Ecofys%20%282013%29%20Implications%20of%20the%20reuse%20of%20captured%20CO2%20-%20report.pdf, checked on 06/03/15.

Hockerts, Kai; Wüstenhagen, Rolf (2010): Greening Goliaths versus emerging Davids — Theorizing about the role of incumbents and new entrants in sustainable entrepreneurship. In *Journal of Business Venturing* 25 (5), pp. 481–492. DOI: 10.1016/j.jbusvent.2009.07.005.

Hueske, Anne-Karen; Guenther, Edeltraud (2015): What hampers innovation? External stakeholders, the organization, groups and individuals: a systematic review of empirical barrier research. In *Manag Rev Q* 65 (2), pp. 113–148. DOI: 10.1007/s11301-014-0109-5.

Naims, H.; Olfe-Kräutlein, B.; Lorente Lafuente, A. M.; Bruhn, T. (2015): CO2-Recycling – Option für Politik und Gesellschaft? Zwölf Thesen zur gesellschaftlichen und politischen Bedeutung von Carbon Capture and Utilisation(CCU)-Technologien. Institute for Advanced Sustainability Studies (IASS). Potsdam (IASS Working Paper).

Strauss, Anselm L.; Corbin, Juliet M. (2015): Basics of Qualitative Research. Techniques and Procedures for Developing Grounded Theory. 4th: Sage publications.

Styring, Peter; Quadrelli, Elsje Alessandra; Armstrong, Katy (2014): Carbon Dioxide Utilisation: Closing the Carbon Cycle. 1st edition: Elsevier Science.

Yin, Robert K. (2013): Case study research: Design and methods. 5th: Sage publications.

471447 Catalytic Dry Methane Reforming for Industrial CO_2 Capture Applications & Syngas Generation

Abstract:
Steam Methane Reforming is the most popular process for the production of H_2 rich syngas from hydrocarbon feed such as natural gas or flare gas in refineries. Generally, the process is operated to obtain a high H_2:CO ratio around 4~7, which is further enhanced via a water gas shift reactor. However, the synthesis of hydrogen gas is accompanied by the generation of a high amount of CO_2 as well, averaging around 9 tonnes of CO_2 per tonne of H_2produced.

Syngas however can be utilized for purposes other than just hydrogen generation, in applications such as the production of synthetic crude oil (via the Fischer-Tropsch process), methanol, ethanol as well as of higher products of enhanced value such as acrylates, organic acids, ether etc. In such processes, the CO molecule is as important as the H_2 molecule. Typically, for most chemical synthesis processes the ratio of consumption of H_2:CO is between 1 â€" 2.5.

Catalytic Dry Methane Reforming[1] is a process for generation of syngas (CO + H_2), utilizing CH_4 and CO_2as feed (1). The reforming reaction is accompanied by the reverse water gas shift reaction (2).

$$CH_4 + CO_2 \rightarrow 2CO + 2H_2 \quad (1)$$

$$CO_2 + H_2 \rightleftharpoons CO + H_2O \quad (2)$$

The product syngas, with a typical composition of H_2:CO â‰¤ 1, can be used to supplement the raw material for the above mentioned processes. The reaction however is highly endothermic, and requires operation at high temperatures (750-900 C). This can potentially lead to significant CO_2 emissions through fuel combustion to provide energy for the reaction. The reaction, however, has been found to consume nearly twice the quantity of CO_2that is produced by combustion of fuel, thus resulting in a carbon capture efficiency of over 50 %. This would make it a great tool to serve the dual purposes of emission reduction as well as valorization of waste greenhouse gases in industry.

Concerns over the increasing global temperatures have triggered an extensive search for viable technologies to mitigate industrial GHG emissions. Implementation of the Dry Methane Reforming process could be a viable solution to this problem for a variety of industries. However, the high temperature operation causes $CH_4$and CO decomposition to form carbon deposits, thus resulting in deactivation of dry reforming catalysts. The lack of a satisfactory catalyst for the DMR system has impeded the application of this process in industry.

Enerkem Inc. is a Canadian company focused on reducing the carbon footprint for cleaner chemical production. The company has developed its technology for converting municipal solid wastes to methanol, which in turn can serve as the base material for secondary chemicals, such as higher alcohols, aldehydes, acrylates etc. It is starting its first commercial operation at the Edmonton Waste Management Centre for the production of methanol. Enerkem Inc. has also developed the E20 dry reforming catalyst, which has been observed to resist deactivation when operated at over 90% CH_4 as well as CO_2conversion for more than 100 hours (Figure 1). A comparison of the emission levels for syngas production using steam methane reforming operating at equilibrium condition and the dry methane reforming using the E20 catalyst have been presented in Table 1.

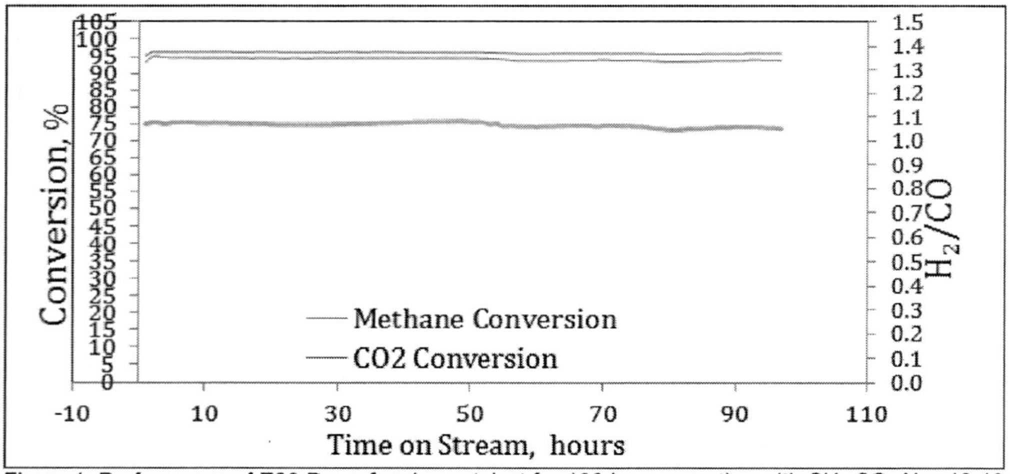

Figure 1. Performance of E20 Dry reforming catalyst for 100 hour operation with CH_4:CO_2:N_2= 40:40:20, reactor temperature of 850 ÌŠC and WHSV of 500 ml/min-g.cat

Kinetic experiments have been conducted on the E20 catalyst and a working kinetic model has been developed. Simulation studies performed on HYSYSÂ® have exhibited the CO_2capture potential of the E20 catalyst when integrated with the synthesis of high value chemicals. To showcase this potential, we have considered the case of acrylic acid synthesis, using syngas as the starting material. The most commonly utilized process for the synthesis of acrylic acid is the partial oxidation of propylene. However, the pathway for synthesis from syngas can be summarized as follows:

Syngas is used to synthesize methanol via reaction 3.

$$CO + 2H_2 \rightarrow CH_3OH \quad (3)$$

Methanol can be utilized to synthesize formaldehyde via the Formox process[3] (4), as well as acetic acid via the methanol carbonylation process[4](5).

$$CH_3OH + \frac{1}{2}O_2 \rightarrow HCHO + H_2O \qquad (4)$$

$$CH_3OH + CO \rightarrow CH_3COOH \qquad (5)$$

The acetic acid and formaldehyde can undergo reaction via an aldol condensation route[5]to form acrylic acid (6).

$$HCHO + CH_3COOH \rightarrow CH_2(OH)CH_2COOH \rightarrow CH_2 = CHCOOH + H_2O \qquad (6)$$

The overall process would allow the valorization of waste CO_2streams into high value acrylate compounds.

	SMR	SMR	DMR	DMR
Reactor Temp (ÌŠC)	1000	1000	850	850
H_2O:CO_2:CH_4	5:0:1	3:0:1	0:1:1	0:1.67:1
H_2:CO	5.5	4.3	0.84	0.71
CH_4 Conversion (%)	99.2	97.2	73.8	67.8
CO_2 Conversion (%)			86.5	54.7
Energy per kg H_2 (kJ)	71933.4	61193.1	81392.6	95834.2
CO_2 produced per kg H_2 (kg)	6.5	5	-9.6	-12.6
CO_2 capture efficiency (%)			68.2	70.5
Energy per kg CO (J)	28242.8	18843.7	4896.3	4841
CO_2 produced per kg CO (kg)	2.5	1.6	-0.58	-0.64
CO_2 capture efficiency (%)			69.1	71.1

Table 1. Comparison of Energy consumption and CO_2emission of Steam Methane Reforming (SMR) and Dry Methane Reforming (DMR) processes

References

1. Bradford, M.; Vannice, M., CO2 reforming of CH4. *Catalysis Reviews* **1999**,41, (1), 1-42.

2. Pakhare, D.; Spivey, J., A review of dry (CO 2) reforming of methane over noble metal catalysts. *Chemical Society Reviews* **2014**,43, (22), 7813-7837.

3. Reuss, G.; Disteldorf, W.; Gamer, A. O.; Hilt, A., Formaldehyde. *Ullmann's Encyclopedia of Industrial Chemistry* **2005**.

4. Yoneda, N.; Kusano, S.; Yasui, M.; Pujado, P.; Wilcher, S., Recent advances in processes and catalysts for the production of acetic acid. *Applied Catalysis A: General* **2001**,221, (1), 253-265.

5. Vitcha, J. F.; Sims, V. A., Vapor Phase Aldol Reaction. Acrylic Acid by Reaction of Acetic Acid and Formaldehyde. *Industrial & Engineering Chemistry Product Research and Development* **1966**, 5, (1), 50-53.

CO_2 as a "Soft" Oxidant for Syngas Production via Chemical Looping

Amey More, Saurabh Bhavsar, Charlie Hansen and Götz Veser

Department of Chemical Engineering, University of Pittsburgh, Pittsburgh, PA

Abstract:

'Chemical Looping Combustion' is a clean combustion technology, which enables fossil fuel combustion with inherent CO_2 capture based on the cyclic oxidation and reduction of an oxygen carrier. While most efforts in chemical looping (CL) are focused on combustion, we previously demonstrated the application of the "CL principle"—the periodic oxidation and reduction of a metal oxide to couple two independent redox reactions—to the activation of CO_2 via reduction to CO [1]. In the present contribution, we investigate and compare CO_2 activation via CL in two different operating modes: In the first scheme, CO_2 reduction is coupled with CH_4 oxidation by using mixtures of Fe and Ni (as alloys or simple physical mixtures) to produce CO and syngas product streams [2]. In the second operating scheme, monometallic Ni carriers are utilized to catalytically crack CH_4, producing pure H_2 streams. The solid carbon deposits are then burnt off with CO_2, overall producing separate CO and H_2 product streams.

Supported metal oxide carriers were synthesized, and a combination of thermo-gravimetric analysis (TGA), X-ray diffraction (XRD), and electron microscopy (TEM) was used to identify structure-reactivity correlations for systematic carrier design. Gas phase conversions and selectivities were determined in periodic fixed bed reactor operation. We find that Fe-Ni alloys can indeed show good activity towards both methane activation and CO_2 reduction, and that the weak oxidant CO_2 allows controlled oxidation of Fe-Ni alloys, which enables selective oxidation of CH_4 to syngas. Remarkably, a simple physical mixture of Fe and Ni far exceeds the reactivity of an equivalent alloy carrier, which can be traced back to a synergistic gas-phase coupling between the two carrier fractions. For the second proposed scheme, we find a strong dependence on the support material, which affects particle size and distribution of the active metal species. Overall, our investigations demonstrate the potential of CO_2 as a "soft" oxidant which enables selective oxidation reactions via novel, intensified chemical looping processes.

[1] M. Najera, et al, *Chem. Eng. Res. Des.* 89 (2011) 1533; Bhavsar, et al, *Chem. Eng. Technol.* 35 (2012) 1281; S. Bhavsar and G. Veser, *Energy Fuels* 27 (2013) 2073; A. More, S. Bhavsar, and G. Veser. *Energy Technology* (2016) AOP.

464674 Process and Material Design for Micro-Encapsulated Ionic Liquids in Post-Combustion CO_2 Capture

Abstract:

Aprotic Heterocyclic Anion (AHA) Ionic Liquids (ILs) have been identified as promising new solvents for post-combustion carbon capture due to their high CO_2 uptake and the high tunability[1,2] of their binding energy with CO_2. Some of these compounds change phase (solid to liquid) on absorption of CO_2; these Phase Change ILs (PCILs)[3] offer the additional advantage that part of the heat needed to desorb the CO_2 from the absorbent is provided by the heat of fusion as the PCIL solidifies upon release of CO_2. However, the relatively high viscosity of AHA ILs and the occurrence of a phase change in PCILs present challenges for conventional absorption equipment. To overcome these challenges we are pursuing the use of new technology to micro-encapsulate the AHA ILs and PCILs. Our partners at Lawrence Livermore National Laboratory have successfully demonstrated this technology in the application of post-combustion carbon capture with sodium and potassium carbonate solutions,[4] and have recently shown the feasibility of micro-encapsulation of an AHA IL for carbon capture.[5] The large effective surface area and high CO_2 permeability of the micro-capsules is expected to offset the drawback of the high IL viscosity and to provide for a more efficient and cost-effective mass transfer operation involving AHA ILs and PCILs.

These opportunities, however, present us with both process and materials design questions. For example, what is the target CO_2 absorption strength (enthalpy of chemical absorption) for the tunable AHA IL? What is the target for micro-capsule diameter in order to obtain a high mass transfer rate and good fluidization performance? What are the appropriate temperatures and pressures for the absorber and stripper? In order to address these and other questions, we have developed a rate-based model of a post-combustion CO_2 capture process using micro-encapsulated ILs. As a performance baseline, we have also developed a rate-based model of a standard packed bed absorber using an un-encapsulated AHA IL absorbent. Using such models we can determine optimal CO_2 capture performance and investigate the sensitivity of the optimum with respect to the key thermo-physical and transport properties of the IL (e.g., CO_2 binding energy, viscosity, etc.) and the micro-capsules (e.g. diameter, CO_2 permeability, etc.). Results of these process and material design studies will be presented, and the performance of this novel micro-encapsulation technology will be assessed.

References

(1) Gurkan, B.; Goodrich, B. F.; Mindrup, E. M.; Ficke, L. E.; Massel, M.; Seo, S.; Senftle, T. P.; Wu, H.; Glaser, M. F.; Shah, J. K.; Maginn, E. J.; Brennecke, J. F.; Schneider, W. F. Molecular Design of High Capacity, Low Viscosity, Chemically Tunable Ionic Liquids for CO_2 Capture. *J. Phys. Chem. Lett.* **2010,** *1*, 3494-3499.

(2) Seo, S.; Quiroz-Guzman, M.; DeSilva, M. A.; Lee, T. B.; Huang, Y.; Goodrich, B. F.; Schneider, W. F.; Brennecke, J. F. Chemically Tunable Ionic Liquids with Aprotic Heterocyclic Anion (AHA) for CO2 Capture. *J. Phys. Chem. B* **2014,** *118*, 5740-5751.

(3) Seo, S.; Simoni, L. D.; Ma, M.; DeSilva, M. A.; Huang Y.; Stadtherr, M. A.; Brennecke, J. F. Phase-Change Ionic Liquids for Postcombustion CO_2 Capture. Energy Fuels, **2014,** *28*, 5968-5977.

(4) Vericella, J. J.; Baker, S. E.; Stolaroff, J. K.; Duoss, E. B.; Hardin, J. O.; Lewicki, J.; Glogowski, E.; Floyd, W. C.; Valdez, C. A.; Smith, W. L.; Satcher, J. H.; Bourcier, W. L.; Spadaccini, C. M.; Lewis, J. A.; Aines, R. D. Encapsulated Liquid Sorbents for Carbon Dioxide Capture, *Nature Commun.*, **2015**, *6*, 6124.

(5) Stolaroff, J. K.; Ye, C.; Oakdale, J. O.; Baker, S. E.; Smith, W. L.; Nguyen, D. T.; Spadaccini, C. M.; Aines, R. D. Microencapsulation of Advanced Solvents for CO_2 Capture. *Faraday Discuss.*, 2016, in press, DOI: 10.1039/C6FD00049E.

457289 Material Development and Assessment of a Thermochemical Energy Storage Based on $CaO/CaCO_3$

Abstract:

The usage of solar energy for heat and power supply is restricted by the fluctuation of solar radiation. To overcome this and to enable a continuous operation of concentrated solar power plants, thermal energy storages units are essential. In this context, reversible thermal decomposition reactions of solids have been proposed and studied for several years [1]: A solid salt is decomposed into another solid and a gaseous component during daytime. The composition reaction, reverse reaction, occurs during night time or at an overcast sky in order to supply continuously energy for a steam cycle. Investigated reaction schemes contain oxides, hydroxides, carbonates and hydrides [2-5]. Due to high storage densities and discharging temperatures between 400 and 1000 °C, the gas-solid reactions are discussed as a promising concept for increasing the efficiency and operation hours of solar thermal power plants.

In previous works [6], the cyclic carbonation/calcination reaction of CaO is proposed as a thermochemical storage for solar power plants. The energy is stored chemically in CaO after the endothermal calcination reaction of $CaCO_3$ (eq. R1). The formed energy-rich CaO and CO_2 are stored separately. The energy is released by initiating the exothermal carbonation reaction (eq. RR1) of CaO and CO_2 at temperatures between 600 and 750 °C.

$$CaCO_3(s) \rightarrow CaO(s) + CO_2(g) \qquad \Delta^R h^{+} = 178 \text{ kJ mol}^{-1} \qquad \text{(R1)}$$

$$CaO(s) + CO_2(g) \rightarrow CaCO_3(s) \qquad \Delta^R h^{+} = - 178 \text{ kJ mol}^{-1} \qquad \text{(RR1)}$$

The involved components are cheap, non-toxic and in case of $CaCO_3$ abundantly available. Nevertheless, a profound challenge is the strong cycle-to-cycle degradation of the material due to sintering. A low cyclic stability implies a decrease of active material after several cycles and thus, the efficiency and storage density is lowered.

Within this study, Al-doped materials are synthesized with different Ca/Al molar ratios in order to stabilize the sorbent structure and prevent the mentioned sintering phenomena. The Ca/Al ratio is varied between 75/25 and 95/5. All samples are experimentally characterized in 20 calcination/carbonation cycles by a thermogravimetric analyzer measuring the CO_2 uptake at predefined conditions [6]. Furthermore, based on the results and a performed enthalpy balance of carbonation and calcination reaction the storage density and the energetic efficiency are calculated in order to evaluate the sorbents out of an energetic point of view. All samples are compared to the performance of the benchmark material (natural CaO).

The experiments show that the formed Ca/Al mixed phase stabilizes the sorbent and the CO_2 uptake behavior can thus be maintained. It could be proven that all synthesized sorbents are characterized by an improved cyclic stability and the relative loss of activity from the first to the 20th cycle is notably reduced. Whereas the benchmark material shows a loss of activity of about 70 % after the 20th cycle, the degradation of the best prepared sorbent is less than 4 %. In context of storage density and efficiency it can be seen that in general, an increase of Al-content entails a lowered maximum achievable storage density and efficiency due to a decreased initial amount of active material (CaO). Nevertheless, by taking the results of the CO_2 uptake measurements into account, all synthesized samples show an improved performance compared to natural CaO in the 20th cycle due to a constant CO_2 uptake. In terms of storage density 2 to 3.5 times higher values can be achieved. For the energetic efficiency a factor of up to 1.2 relative to the benchmark material is obtained.

Thus, based on these experimental and theoretical investigations it can be confirmed that the addition of a certain amount of initial inert material is justified in order to optimize the performance of the material. The results of the study lay the foundation for the usage of the $CaO/CaCO_3$ reaction couple as a high efficient thermochemical heat storage.

[1] W. Wentworth, E. Chen, Solar Energy, 18 (1976), 205-214

[2] A. J. Carrillo, J. Moya, A. Bayón, P. Jana, A. Victor, M. Romero, J. Gonzalez-Aguilar, D. Serrano, P. Pizarro, J.M. Coronado, Solar Energy Materials and Solar Cells, 123 (2014), 47-57

[3] F. Schaube, A. Wörner, R. Tamme, Journal of Solar Energy Engineering, 133 (2011), 1006-1012

[4] M. Aihara, K. Tanaka, M. Watanabe, T. Takeuchi, H. Habuka, Journal of Chemical Engineering of Japan, 40 (2007), 1270-1274

[5] M. Paskevicius, D. Sheppard, K. Williamson, C. Buckley, Energy, 88 (2015), 469-477

[6] J. Obermeier, B. Müller, K. Müller, W. Arlt, Energy Technology, 4 (2016), 123 – 135

AUTHOR INDEX

AUTHOR INDEX

AUTHOR INDEX

AUTHOR INDEX